Springer Theses

Recognizing Outstanding Ph.D. Research

T0272069

T0324747

Aims and Scope

The series "Springer Theses" brings together a selection of the very best Ph.D. theses from around the world and across the physical sciences. Nominated and endorsed by two recognized specialists, each published volume has been selected for its scientific excellence and the high impact of its contents for the pertinent field of research. For greater accessibility to non-specialists, the published versions include an extended introduction, as well as a foreword by the student's supervisor explaining the special relevance of the work for the field. As a whole, the series will provide a valuable resource both for newcomers to the research fields described, and for other scientists seeking detailed background information on special questions. Finally, it provides an accredited documentation of the valuable contributions made by today's younger generation of scientists.

Theses are accepted into the series by invited nomination only and must fulfill all of the following criteria

- They must be written in good English.
- The topic should fall within the confines of Chemistry, Physics, Earth Sciences, Engineering and related interdisciplinary fields such as Materials, Nanoscience, Chemical Engineering, Complex Systems and Biophysics.
- The work reported in the thesis must represent a significant scientific advance.
- If the thesis includes previously published material, permission to reproduce this must be gained from the respective copyright holder.
- They must have been examined and passed during the 12 months prior to nomination.
- Each thesis should include a foreword by the supervisor outlining the significance of its content.
- The theses should have a clearly defined structure including an introduction accessible to scientists not expert in that particular field.

More information about this series at http://www.springer.com/series/8790

Kwai Hei Li

Nanostructuring for Nitride Light-Emitting Diodes and Optical Cavities

Doctoral Thesis accepted by
The University of Hong Kong, China

 Springer

Author
Dr. Kwai Hei Li
Department of Electrical and Electronic
Engineering
The University of Hong Kong
Hong Kong
China

Supervisor
Prof. Anthony H.W. Choi
Department of Electrical and Electronic
Engineering
The University of Hong Kong
Hong Kong
China

ISSN 2190-5053   ISSN 2190-5061   (electronic)
Springer Theses
ISBN 978-3-662-48607-8   ISBN 978-3-662-48609-2   (eBook)
DOI 10.1007/978-3-662-48609-2

Library of Congress Control Number: 2015952029

Springer Heidelberg New York Dordrecht London

Printed on acid-free paper

Springer-Verlag GmbH Berlin Heidelberg is part of Springer Science+Business Media
(www.springer.com)

Parts of this thesis have been published in the following journal articles:

1. Li KH, Feng C, Choi HW (2014) Analysis of micro-lens integrated flip-chip InGaN light-emitting diodes by confocal microscopy. Appl Phys Lett 104 (5). doi: Artn 051107 Doi 10.1063/1.4863925
2. Li KH, Zang KY, Chua SJ, Choi HW (2013) III-nitride light-emitting diode with embedded photonic crystals. Appl Phys Lett 102 (18). doi: Artn 181117 Doi 10.1063/1.4804678
3. Li KH, Cheung YF, Zhang Q, Choi HW (2013) Optical and Thermal Analyses of Thin-Film Hexagonal Micro-Mesh Light-Emitting Diodes. Ieee Photonic Tech L 25 (4):374–377 Doi 10.1109/Lpt.2013.2238621
4. Li KH, Ma ZT, Choi HW (2012) Tunable clover-shaped GaN photonic bandgap structures patterned by dual-step nanosphere lithography. Appl Phys Lett 100 (14). doi: Artn 141101 Doi 10.1063/1.3698392
5. Li KH, Ma ZT, Choi HW (2012) Single-mode whispering gallery lasing from metal-clad GaN nanopillars. Opt Lett 37 (3):374–376
6. Li KH, Choi HW (2011) InGaN light-emitting diodes with indium-tin-oxide photonic crystal current-spreading layer. J Appl Phys 110 (5). doi: Artn 053104 Doi 10.1063/1.3631797
7. Li KH, Ma ZT, Choi HW (2011) High-Q whispering-gallery mode lasing from nanosphere-patterned GaN nanoring arrays. Appl Phys Lett 98 (7). doi: Artn 071106 Doi 10.1063/1.3556281
8. Li KH, Choi HW (2011) Air-spaced GaN nanopillar photonic band gap structures patterned by nanosphere lithography. J Appl Phys 109 (2). doi: Artn 023107 Doi 10.1063/1.3531972

Supervisor's Foreword

This thesis describes the scientific achievements of Dr. Kwai Hei Li, which were made during his doctoral program at the Department of Electrical and Electronic Engineering, the University of Hong Kong. His research work is centered around the nanostructuring of gallium nitride optical materials and devices using a patterning technique that he has developed and mastered. Taking advantage of the natural abilities of spherical particles to self-assemble into regular close-packed patterns, Dr. Li applied nanosphere lithography (NSL) to demonstrate a range of novel nanostructures on nitride semiconductors. These regularly patterned structures are of great technological importance, enabling the manipulation light at subwavelength scales through nanophotonic effects. Compared with other top-down approaches for fabricating III-nitride nanostructures such as e-beam lithography, NSL allows large-area patterning. NSL also overcomes resolution issues arising from the optical diffraction limit in optical lithography. Low setup costs and high throughput make NSL potentially suitable for mass production, such that the technologies developed can readily be incorporated into products.

Dr. Li has applied NSL to nitride-based light-emitting diodes (LEDs) and laser structures, and has successfully achieved significant accomplishments. In one of his earlier works, he textured the surface of LEDs via NSL to form regular close-packed nanopillars and nanolenses, which has been proven to enhance light extraction, as well as reducing the emission divergence, making LEDs more efficient and their beams more directional. He further extended the use of NSL to pattern novel photonic bandgap structures, namely air-spaced nanopatterns and clover-shaped structures. These are extremely useful novel photonic structures for suppressing the propagation of lateral photons so as to enhance the light extraction from the top surface of LEDs. His investigations led to the possibility of increasing the light extraction efficiency and modifying emission characteristics of LEDs via photonic bandgap using an easily achievable nanostructuring technique. For the work on laser structure, he has achieved novel results by modifying the regular NSL process, turning it into dual-step NSL to form clover-like structures. Additionally he has designed and fabricated a well-ordered array of nanoring resonators with high Q factor and low lasing threshold. He also further demonstrated single-mode lasing from metal-clad nanopillar structure. Compared with other fabrication techniques, NSL is superior in producing high-density well-defined nanocavities.

Adopting nonconventional approaches in research is always deemed to be risky as results are not guaranteed. However, Dr. Li not in the least concerned about overcoming multiple obstacles along the way through relentless hard work and perseverance. This excellent thesis is a collection of his research outputs that were previously published in various scientific journals and conference papers. Without a doubt, his research works open up new perspectives on tackling some of the major issues in the field of nitride optoelectronics.

Hong Kong   Prof. Anthony H.W. Choi
June 2015

Contents

Abbreviations

2-D	Two-dimensional
3-D	Three-dimensional
AFM	Atomic force microscopy
Ag	Silver
Al	Aluminum
ALD	Atomic layer deposition
AlN	Aluminum nitride
Ar	Argon
Au	Gold
BCl_3	Boron trichloride
CCD	Charge-coupled device
CHF_3	Trifluoromethane
Cl_2	Chlorine
CP	Close-packed
Cp_2Mg	Bis cyclopentadienyl magnesium
Cu	Copper
CW	Continuous Wave
DBR	Diffracted Bragg reflector
dc	Direct current
DI	Deionized water
DPSS	Diode-pumped solid state
EBL	Electron beam lithography
EL	Electroluminescence
ELO	Epitaxial lateral overgrowth
EQE	External quantum efficiency
F–B	Fabry–Pérot
FCC	Face-centered cubic
FDTD	Finite-difference time domain
FE-SEM	Field emission scanning electron microscope
FIB	Focused ion beam
FWHM	Full-width at half-maximum

GaN	Gallium nitride
H_2	Hydrogen
HCl	Hydrochloric acid
HCP	Hexagonally close-packed
He	Helium
He-Cd	Helium cadmium
HF	Hydrofluoric acid
ICP	Inductively coupled plasma
InGaN	Indium gallium nitride
InN	Indium nitride
IPA	Isopropyl alcohol
IQE	Internal quantum efficiency
IR	Infrared
ITO	Indium tin oxide
I–V	Current–voltage
Lat-μLED	Lateral micro-LED
Lat-LED	Lateral LED
LED	Light-emitting diode
LEE	Light extraction efficiency
L-I	Light output-current
LLO	Laser lift-off
LWIR	Long wavelength infrared
MOCVD	Metalorganic chemical vapor deposition
MQW	Multi-quantum well
N.A.	Numerical aperture
N_2	Nitrogen
NCP	Non-close-packed
NH_3	Ammonia
Ni	Nickel
NSL	Nanosphere lithography
O_2	Oxygen
OLED	Organic light-emitting diode
PBG	Photonic bandgap
PhC	Photonic crystal
PL	Photoluminescence
PWE	Plane wave expansion
QW	Quantum well
RF	Radio frequency
RIE	Reactive-ion etching
RMS	Root-mean-squared
RT	Room temperature
RWCA	Rigorous coupled-wave analysis
SCCM	Standard cubic centimeters per minute
SDS	Sodium dodecyl sulfate
SEM	Scanning electron microscope

SF$_6$	Sulfur hexafluoride
SiO$_2$	Silicon dioxide
TE	Transverse electric
TF-μLED	Thin-film micro-LED
TF-LED	Thin-film LED
Ti	Titanium
TIR	Total internal reflection
TM	Transverse magnetic
TMGa	Trimethylgallium
UV	Ultraviolet
WG	Whispering gallery

Chapter 1
Introduction

1.1 III–V Nitrides

Remarkable advancements in group III-nitride semiconductors have made possible direct-band gap emission throughout the ultraviolet to visible spectral bands, hence the emergence of an entirely new range of light emitters. As illustrated in Fig. 1.1, the possibility of alloying the III-nitrides to form ternary ($Al_xGa_{1-x}N$, $In_xGa_{1-x}N$, and $In_xAl_{1-x}N$) and quaternary ($In_xAl_yGa_{1-x-y}N$) alloys makes them able to cover a wide spectral range from 1.9 eV (InN) to 6.2 eV (AlN). The expression for the energy band gap of the ternary alloys $A_xB_{1-x}N$, where A and B are the metallic components (In, Al or Ga), can be generally written as

$$E_g^{A_xB_{1-x}N} = xE_g^{AN} + (1-x)E_g^{BN} - b_{AB}(1-x) \tag{1.1}$$

while the energy band gap of the $In_xAl_yGa_{1-x-y}N$ quaternary alloy is depicted as a function of concentrations x and y:

$$\begin{aligned}E_g(x, y) = {} & (1 - x - y)E_g^{GaN} + E_g^{AlN} + yE_g^{InN} - b_{AlGaN}x(1 - x) \\ & - b_{InGaN}y(1 - y) - (b_{AlInN} - b_{AlGaN} - b_{InGaN})xy\end{aligned} \tag{1.2}$$

where the bs are the bowing coefficients of respective alloys. Moreover, owing to its intrinsically attractive features of mechanical and thermal stabilities, III–V nitride semiconductor has been considered as the leading material for optoelectronic applications [1–3].

In the III-nitride material system, the face-centered cubic (FCC) zincblende and the hexagonal close-packed (HCP) wurtzite are the common crystal structures. The zincblende structure is formed from cubic unit cells and provides a higher degree of crystallographic symmetry due to its identical lattice constants along three perpendicular directions. The wurtzite structure consists of hexagonal unit cells with

© Springer-Verlag Berlin Heidelberg 2016
K.H. Li, *Nanostructuring for Nitride Light-Emitting Diodes and Optical Cavities*, Springer Theses, DOI 10.1007/978-3-662-48609-2_1

Fig. 1.1 Band gap energy
versus lattice constant of
III-nitride semiconductors

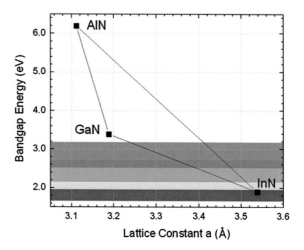

Table 1.1 Lattice parameters
for AlN, GaN, and InN
materials

Lattice parameter	AlN	GaN	InN
a (Å)	3.111	3.189	3.534
c (Å)	4.978	5.185	5.1718

two different lattice constants. With its thermodynamic stability at ambient conditions, the wurtzite crystalline is mostly adopted in optoelectronic applications and the corresponding lattice parameters of AlN, GaN, and InN are listed in Table 1.1, where the lattice parameters a and c are defined as the edge length along the hexagonal basal plane and the spacing in between two identical hexagonal lattice planes respectively.

1.2   Motivations

Undoubtedly, the development of nitride-based light-emitting diodes (LEDs) and laser diodes (LDs) represented a quantum leap in the advancement of optoelectronics. Although GaN LEDs are on the way to replace lightbulbs and fluorescent tubes, most of the emitted light suffers from total internal reflection and is trapped inside the device, resulting in extremely low light extraction efficiency. Short-wavelength lasers offer great impact to optical high-density storage and displays, but there is a lack of practical approaches to construct laser cavities for short-wavelength emission. Thus, the development of processes feasible and suitable for mass production of nanoscale features for III-nitride optoelectronic devices is definitely a task of high priority. The research works in this dissertation are based on novel nanostructuring technique—nanosphere lithography (NSL) to pattern various nanostructures on nitride semiconductors. These regularly-patterned structures are of great technological importance, enabling us to manipulate light at subwavelength dimensions through nanophotonic effects. Compared with other possible top-down approaches for fabricating III-nitride nanostructures, such as electron-beam evaporation and focused ion beam, NSL has its predominance in efficiently forming arrays across large areas. Moreover, NSL overcomes resolution issues arising from diffraction limit in optical lithography and even limitation of beam spot in electron-beam lithography.

The incorporation of two-dimensional (2-D) PhCs onto the surfaces of nitride-based LEDs can be considered as one of the most effective ways to promote light extraction efficiency. Photonic crystal in space with a periodicity comparable to the emission wavelength offers the unique ability of manipulating spontaneous emission of LED, which is extremely useful for extracting guided modes to air and thus enlarging the escape cone. In spite of all its attractiveness and promises, a lack of practical approach toward fabrication of regularly periodic nanoscale patterns is still a major challenge. Ordered hexagonal nanostructures patterned by NSL open an opportunity and become a breakthrough to realize the mass production of photonic crystal structures on the light emitters.

Although diode-pumped solid-state (DPSS) lasers are now widely available even for short-wavelength lasing, the laser configuration involves crystal-based nonlinear frequency up-conversion, a process which is complex and energy inefficient. Undoubtedly, the long-term prospect is the elimination of the need for frequency up-conversion by adopting nitride semiconductors for lasing applications, which offers many advantages including energy-efficiency, low heat generation, simple fabrication, and assembly processes as well as miniature footprints. Among all possible lasing mechanisms in nitride-based material, whispering gallery mode based on cylindrical dielectric resonator is a promising candidate with attractive properties of intrinsically high Q factor, low lasing threshold, and simple fabrication process. Such circular cavities are presently restricted to proof-of-concept small-scale demonstrations in laboratories. Again, with the aid of NSL, the high flexibility of nanosphere dimensions offers a viable and effective solution for generating high-finesse optical cavity for photon confinement in a spectral range of visible wavelength and even ultraviolet radiation.

1.3 Objective and Organization of This Dissertation

This thesis presents a cost-effective practical approach to tackle two major concerns in the field of nitride optoelectronics; low light extraction efficiency of LEDs and lack of a practical approach toward fabrication of high-quality short-wavelength cavities. Diversely functional nanostructures are fabricated to boost the optical performance of LEDs and optical cavities. This thesis consists of seven chapters. This chapter gives a basic overview of III-nitride semiconductor and also outlines the objective and research motivations.

Chapter 2 provides an introduction to LED efficiency and the basic theories of optical micro-cavities. Several strategies for improving light extraction efficiency of LEDs are described. Lasing mechanisms of three common micro-cavities are briefly discussed. This chapter also gives a detailed description of nanosphere lithography.

In Chap. 3, sphere-patterned structures functioning as two-dimensional photonic crystals are applied to GaN LED materials for realizing enhancement of light extraction. In the first part, two EL devices incorporating close-packed nanopillar arrays are reported. In the latter part, two photonic band gap structures, namely air-spaced and clover-shaped structures are fabricated by modified nanosphere lithography to extend the function of sphere-patterned structures.

In Chap. 4, the sphere-patterned structures act as whispering gallery mode cavities, which support resonant mode in the strong confinement regime. Two whispering gallery mode cavities, namely nano-ring array and metal-clad nanopillar array, are reported, both of which are fabricated via modified nanosphere lithography. Optical pumped blue/violet lasing at room temperature is observed from both cavities. FDTD simulations are employed to analyze the optical performance of lasing cavities.

Chapter 5 mentions a hexagonally close-packed 1-μm lens array monolithically patterned by nanosphere lithography. Morphology, optical, and electrical properties of patterned flipchip devices are provided. The focusing behavior of close-packed lens array is investigated by confocal microscope and the angular divergence of microlens-integrated LED is found to be significantly reduced, as verified by angular-resolved EL measurement.

In Chap. 6, a thin-film LED with interconnected micromesh array is introduced. Optical, electrical, and thermal properties of vertical micro-LED are studied, compared with conventional lateral LED. Various simulations are conducted to characterize the optical and thermal performance of thin-film devices.

Chapter 7 summarizes the works presented in this thesis and suggests some possible future research works.

References

1. Akasaki I, Sota S, Sakai H, Tanaka T, Koike M, Amano H (1996) Shortest wavelength semiconductor laser diode. Electr Lett 32 (12):1105–1106. doi:10.1049/El:19960743
2. Nakamura S, Senoh M, Nagahama S, Iwasa N, Yamada T, Matsushita T, Kiyoku H, Sugimoto Y (1996) InGaN-based multi-quantum-well-structure laser diodes. Jpn J Appl Phys 2 35 (1B):L74–L76. doi:10.1143/Jjap.35.L74
3. Ponce FA, Bour DP (1997) NItride-based semiconductors for blue and green light-emitting devices. Nature 386 (6623):351–359. doi:10.1038/386351a0

Chapter 2
Background

2.1 LED Efficiency

It is well-known that the performance of a LED is governed by the wall-plug efficiency defined as the ratio of total optical output power from emitter to the electrical input power. It indicates how efficiently the electrical power can be converted into optical power and mathematically expressed as the product of four factors, electrical efficiency, injection efficiency, internal quantum efficiency, and light extraction efficiency.

$$\eta_{\text{wall}} = \eta_{\text{ele}} \cdot \eta_{\text{EQE}} = \eta_{\text{ele}} \cdot \eta_{\text{inj}} \cdot \eta_{\text{int}} \cdot \eta_{\text{extraction}} \tag{2.1}$$

The electrical efficiency is primarily limited by ohmic losses and driver losses and shows how the energy can be acquired from the power source to drive an operating LED. The product of remaining parts is regarded as the external quantum efficiency (EQE) which indicates the portion of photons emitted from the devices injected carrier.

2.1.1 Injection Efficiency

Prior to electron-hole recombination, an injection of electron-hole pair into the active layer are necessitated. The injection efficiency is defined as the proportion of injected electrons being able to reach the active region. The injection efficiency can be expressed as:

$$\eta_{\text{inj}} = \frac{J_p}{J_p + J_n + J_s} \tag{2.2}$$

© Springer-Verlag Berlin Heidelberg 2016
K.H. Li, *Nanostructuring for Nitride Light-Emitting Diodes and Optical Cavities*, Springer Theses, DOI 10.1007/978-3-662-48609-2_2

where J_p is the minority carrier hole diffusion current, J_n is the minority carrier electron diffusion current, and J_s is the space charge recombination current in p-n junction.

2.1.2 Internal Quantum Efficiency

During radiative recombination, an electron from the conduction band directly combines with a hole in the valence band to generate a photon. However, the non-radiative recombination indeed competes with radiative recombination under practical conditions. Such unwanted processes occur via the Auger processes or the defect levels to produce heat or phonons [1]. Internal quantum efficiency (IQE) is the fraction of electron–hole pairs in the active region that can recombine radiatively to produce photons.

$$\eta_{\text{int}} = \frac{P_{\text{int}}/(hv)}{I/e} \tag{2.3}$$

where P is the optical power emitted from the active region.

2.1.3 Light Extraction Efficiency

In an ideal scenario, all photons generated at the active region are expected to escape from the LED and enter into free space, achieving unity extraction efficiency. However, in real situation, emitted light may be partially reabsorbed by the LED substrates, active layer, semi-transparent current spreading film, metallic contacts, bonding wires, etc. The light extraction efficiency (LEE) is typically defined as the portion of light generated in the active region that can emit into free space

$$\eta_{\text{ext}} = \frac{P/(hv)}{P_{\text{int}}/(hv)} \tag{2.4}$$

where P is the optical power emitted from the active region into free space.

2.2 Strategies for Light Extraction

The growing demand for blue light LEDs has also prompted for devices with maximal external quantum efficiency (EQE), which is determined by both internal quantum efficiency (IQE) and light extraction efficiency (LEE). With the rapid and massive improvements of growth techniques, epitaxial structures and crystal quality, the IQE has been greatly enhanced to more than 80 % [2]. However, the extremely low extraction efficiency (<10 %) is still one of the major bottlenecks restricting the performance of LEDs [3], attributed to absorption of substrate, current spreading layer, ohmic contacts and bonding wire, as well as the main challenge of total internal reflection, thus implying that there is still plenty of room for improving the LEE. In the following parts, the influence of total internal reflection is discussed. Numerous approaches aiming to extract optically guided light from devices and suppress total internal reflection are highlighted, including surface roughening, microLEDs, geometrical shaping, and photonic crystal. These methods rely on the formation of nonparallel surfaces to minimize reflections and reduce reabsorption loss, albeit at different dimensional scales.

2.2.1 Total Internal Reflection

Owing to high refractive index contrast at semiconductor/air interface, majority of photons emitted from the active region are remained trapped. The light trapping phenomenon is known as total internal reflection (TIR), strictly limiting the light extraction efficiency of LED (Fig. 2.1). According to Snell's law, TIR occurs when light rays strike on the flat-top semiconductor/air interface with incident angle greater than critical angle

$$\theta_c = \sin^{-1}\left(\frac{n_{\text{air}}}{n_{\text{sc}}}\right) \tag{2.5}$$

where n_{air} and n_{sc} are the refractive indexes of air and semiconductor. Photons outside of the escape angle are likely to be reabsorbed after multiple reflections. In particular, the refractive index is about 2.45 for III-nitride semiconductor and the light extraction angle (escape angle or escape cone) is about 23.5°. TIR resulting from the narrow escape cone prevents the photons from escaping from the semiconductor. Moreover, the light extraction efficiency of the top emission surface can be roughly estimated as

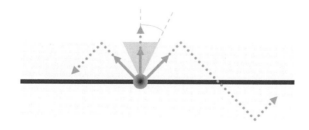

Fig. 2.1 Depicts how the emitted light remains trapped within the device

$$\eta_{\text{ext}} = \int_0^{\theta_c} T_f(\theta)\alpha(\theta, d) \sin\theta d\theta \tag{2.6}$$

where $T_f(\theta)$ is the Fresnel transmission coefficient function and $\alpha(\theta,d)$ is the absorption coefficient function. The flat-top emission surface of the conventional LEDs is found to be as low as 4 % [4] while the overall extraction efficiency is strictly limited to around 12 %.

2.2.2 Surface Roughing

A popular, cost effective and practical approach is to roughen the surface of LED chip via natural chemical etching. Common roughening techniques, including photo-electrochemical chemical etching and wet etching, are capable of developing high-density randomly oriented miniature facets/features on the LED surface [5–8]. The processed surface can randomize the path of trapping and significantly increase the probability for light striking the boundary at an angle close to normal. Surface roughing technique can possibly produce about a factor of enhancement in the light output power and effectively scatter the trapped light outside the LED devices (Fig. 2.2).

2.2.3 MicroLEDs

When emitted, light incidents upon the boundary at angle an greater than critical angle; it suffers total internal reflection and become laterally guided modes, which

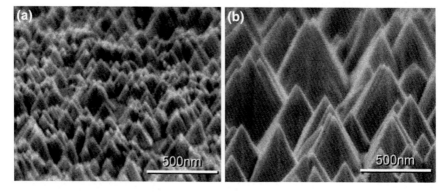

Fig. 2.2 Roughened surface of GaN surface etched by PEC Method. Reprinted from Ref. [5] with permission from AIP Publishing LLC

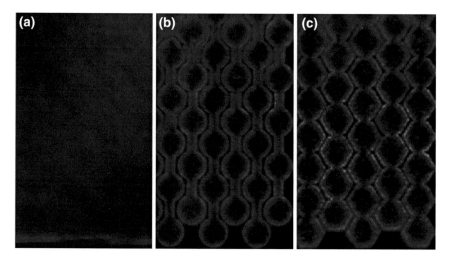

Fig. 2.3 Optical microphotographs showing emission regions of the **a** large-area, **b** microdisk, and **c** microhexagon LEDs. Reprinted from Ref. [9] with permission from AIP Publishing LLC

can either be reabsorbed or escaped from the edges of LED. To increase the changes for light to be extracted into free space before reabsorption, microLEDs provide additional photon escape pathways through the peripheries of microstructures [9–11]. The interconnected microstructures are generally formed by dry etching process so as to remove the material between the microelements, thus significantly increasing the exposed sidewalls [12]. The enhancement in light extraction is attributed to the increased surface area especially the etched sidewall and the CCD images shown in Fig. 2.3, clearly indicate that higher brightness is observed along the edges of microstructures.

2.2.4 Geometrical Shaping

For a conventional LED with cubical geometry, a light ray reflected from one face is likely to hit another parallel facet and bounce around inside the LED chip until it is reabsorbed. One way to overcome this problem is to change the shape of the LED die by creating beveled sidewalls such that the facet pairs are no longer parallel and possibly alter its propagation direction of reflected light [13, 14]. C.C. Kao et al. demonstrated the light output power of a nitride-based LED with 22° undercut sidewall LED was enhanced by 70 % [15]. Moreover, Wang et al. reported various polygonal LEDs shaped with laser micromachining (Fig. 2.4) and proved that LEDs with polygonal geometries increases the light extraction compared to conventional rectangle LEDs [16].

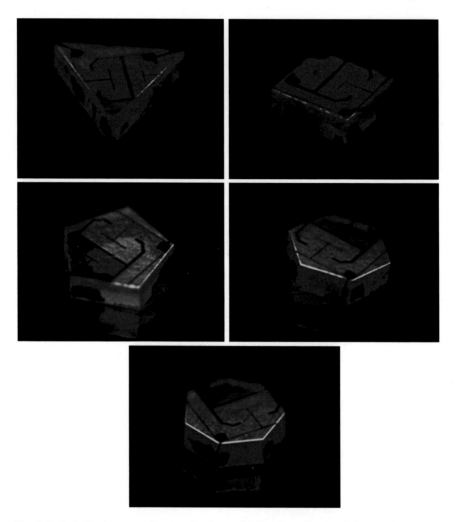

Fig. 2.4 Optical microscopy images of polygonal LED chips. Reprinted from Ref. [16] with permission from AIP Publishing LLC

2.2.5   Photonic Crystal

Photonic crystals (PhCs) [17, 18], with unique capabilities of being able to control and manipulate the propagation of light, have been widely adopted in diverse optoelectronic and photonic applications, including laser resonant cavities [19], high-speed optical fiber transmission [20], and polarization filtering [21]. The incorporation of 2-D PhCs onto the surfaces of nitride-based LEDs enable strong interaction of the guided modes and has also been demonstrated to effectively promote light extraction efficiency [22, 23]. Such ordered periodic nanostructures (Fig. 2.5),

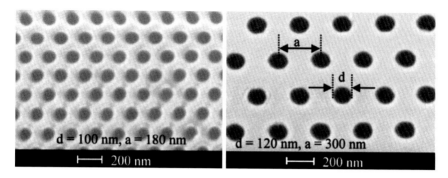

Fig. 2.5 Hexagonal air-hole array patterned by electron-beam lithography (*right*). Reprinted from Ref. [18] with permission from AIP Publishing LLC

with the ability of manipulating spontaneous emission, can be extremely useful for extracting guided modes to air, thus enlarging the escape cone.

The light extraction behaviors of 2-D PhC can be explained by the dispersion diagrams showing normalized frequency versus in-plane wave vector. As illustrated in Fig. 2.6, the green solid line in the band structure, namely the light line, represents a dividing line between guided and leaky modes. With the presence of PhC, band folding will occur at the Brillouin zone edges and guided modes are folded above the air light line, meaning that the guided modes can radiate out form the device. The radiative modes located at the region above the light line corresponds to leaky modes in which the optical mode leaks energy into the surrounding air as it propagates down the waveguide. For the frequency bands that are below the light line, they are the guided modes and do not leak energy as they propagate. Therefore, the light extraction enhancement originates from the coupling of leaky modes above the light line of the band structure. In addition, PhC can also act as

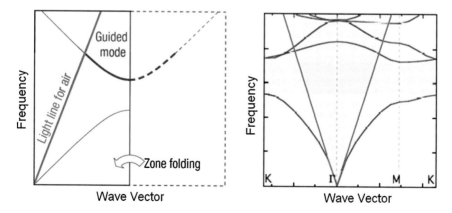

Fig. 2.6 Band diagram illustrating the band folding effect (*left*) and band structure of well-designed PhC (*right*)

2-D diffraction gratings in slabs to extract guided modes to the air and to redirect the emission directions.

With a well-defined periodic arrangement and with sufficiently large refractive index contrast between GaN and ambient, a photonic bandgap (PBG) may be established, which forbids the propagation of light within a specific range of frequencies dependent on the dimension and pitch of the array, as illustrated in Fig. 2.6. The PBG can thus be exploited for suppressing lateral wave-guiding and possibly redirect a significant proportion of trapped photons for extraction, overcoming one of the major limitations of nitride LEDs.

2.3 Lasing Mechanisms

Generally, an optical cavity or resonator, which comprises of two or more mirrors, is capable of confining and storing light at certain resonance frequencies such that a standing wave is established within the cavity. Various optical resonator configurations are depicted in Fig. 2.7. The most common one, so called Fabry–Pérot (F-B) resonator [24, 25], comprises two parallel planar mirrors and forces the light to bounce back and forth within them. Those mirrors can be formed by periodic dielectric mirror such as distributed Bragg reflector (DBR) to confine light with desired frequency in a particular direction. The resonant mode occurs when an integral multiple of half-wavelengths fit into the cavity spacing of length.

Whispering-gallery (WG) mode [26, 27] is mostly constructed within the circular, hemispherical, or even elliptical cavities. When light is propagating around the edge of cavity, it will be totally reflected at boundary of resonator and confined a closed circular path. Thereby, the confined ray propagates around the inside rim of a resonator with its incident angle greater than the critical angle. In order to obtain constructive interference, the propagation path per cycle, which is approximately equal to the circumference of resonator, should be an integer multiple of wavelength of light.

Two-dimensional (2-D) photonic crystals (PhCs) with defects are also used to make microcavities [28, 29]. The periodic dielectric structure exhibits photonic bandgaps which are able to suppress the propagation of light with the bandgap frequency ranges. A defect in the 2-D periodic PhCs can be designed as a missing element in an air-hole array. For the wavelengths overlapping with the PBG, the periodic structure surrounding the defect acts as the reflector, so that light is trapped within the defect. The defect is then regarded as a microcavity resonator and possibly provides extremely small mode volume and high Q-factor exceeding 10^5, but the main challenge for such cavities are the difficulty for its precise design and fabrication.

Those optical resonators are the common light-confinement mechanisms adopted for semiconductor lasers. The optical path lengths of such cavities, dependent on the vertical heights of F–P cavities, or lateral dimensions of WG circular cavities and photonic bandgap (PBG) structures, are crucial factors determining the lasing

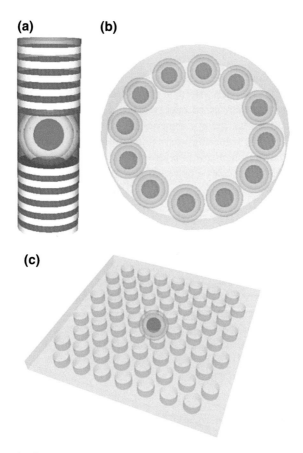

Fig. 2.7 Schematic diagrams depicting three common light-confinement mechanisms: **a** Fabry–Pérot, **b** whispering-gallery, and **c** 2-D photonic crystal defect cavities

characteristics. To confine propagating photons in controllable and predictable manner, these optical cavities are usually designed to be of the order of the wavelength or even at subwavelength scales.

2.4 Quality Factor and Loss of a Resonant Cavity

The quality factor (Q factor) is a key figure of merit for resonator and defined as the fraction of the energy stored in a cavity to the energy dissipated per cycle. For example, high value of Q factor corresponds to a low rate of energy loss relative to the stored energy of a resonator. The value of Q factor can be simply calculated by measuring the spectrum of the resonant mode

$$Q = \frac{\lambda}{\Delta\lambda} \tag{2.7}$$

where λ is the resonant wavelength, and $\Delta\lambda$ is the full-width-at-half-maximum (FWHM) of the resonant peak. An ideal cavity/resonator offers an infinite value of Q factor, meaning that there is no energy dissipated per cycle. In practical applications, a resonant cavity always experiences the energy losses comprising of material absorption loss, scattering loss, radiation loss, and coupling loss. The sum of the first three loss mechanisms is referred to as the intrinsic loss of a resonator while the overall Q factor can be determined by the individual loss terms:

$$\frac{1}{Q} = \frac{1}{Q_{\text{abs}}} + \frac{1}{Q_S} + \frac{1}{Q_{\text{rad}}} + \frac{1}{Q_C} \tag{2.8}$$

where Q_{abs} corresponds to the light absorption of the material; Q_S is related to the scattering loss due to surface inhomogeneities and contaminants; Q_{rad} is the radiation loss originated from the escape of light energy through a curvature surface; and Q_C is the loss induced by a waveguide coupling. Scattering loss can be reduced by improving the fabrication process to minimize the resonator surface roughness. Radiation loss strongly depends on the size of a resonator and can be reduced by increasing the cavity size. It is also worth noting that when the resonator size is larger than some certain dimension, the radiation loss will become negligible compared to other loss mechanisms.

2.5   Nanosphere Lithography

As the characteristic length scales of PhCs structures and the dimension of resonators are of the order of the wavelength, nanopatterning techniques are involved during fabrication of short-wavelength optoelectronic devices, often increasing manufacturing costs. Unlike conventional AlInGaP emitters which can be processed by standard microlithographic techniques, the blue/UV PhC and cavities require further dimensional shrinkage in order to fulfill constraints associated with short-wavelength light interaction, so that traditional optical patterning techniques are no longer able to offer the required resolutions due to the diffraction limit. While direct-write techniques such as electron-beam lithography (EBL) [18, 30] and focused ion beam (FIB) milling [31, 32] are capable of producing arbitrary 2-D feature accurately down to the nanometer scale, they also each have their own drawbacks. The nanofabrication methods are summarized in Table 2.1. High equipment cost and time-consuming point-by-point processing and thus low throughput make large-volume manufacturing impractical. To overcome such limitations, a high-throughput yet low-cost approach is introduced that is particularly suitable for the processing of hard nitride semiconductors: nanosphere lithography (NSL).

Table 2.1 Comparison of nanofabrication methods

	EBL	FIB	NSL
Etch selectivity	Low	N/A	High
Equipment cost	High	High	Low
Preparation time	Long	Long	Short

2.5.1 Overview

Nanosphere lithography (NSL) [33] is an inexpensive and ultra-efficient nanopatterning technique with attractive abilities of producing well-ordered periodic arrays over large areas with minimal processing time. With wide ranges of commercially-available sphere dimensions from hundreds of micrometer down to tens of nanometers, the spheres can self-assemble into mono- and multi-layers of periodically ordered array (Fig. 2.8). Nanospheres have previously found uses in technologies such as catalysis, biochemical devices, cell cultures, surface-enhanced Raman spectroscopy, and sensing application. In the field of optoelectronics, ordered periodic structures based on nanospheres may be used as PhCs to manipulate the flow of light; in particular, 3-D PhCs [34–36] which can interact with light in both the vertical and lateral directions. In this chapter, we shall mainly focus on lithographic processes for generating 2-D monolayers of silica spheres which serve as high etch selectivity masks for GaN materials during dry etch, with the target of producing various 2-D regular nanostructure arrays on the surface of the wafer. We shall first discuss the fundamental steps in the formation of self-assembled ordered monolayer in hexagonal-close-packed (HCP) arrangement for pattern transfer.

2.5.2 Process Development

The formation of an ordered HCP monolayer over large areas on a substrate is commonly achieved via one of the three coating strategies including vertical

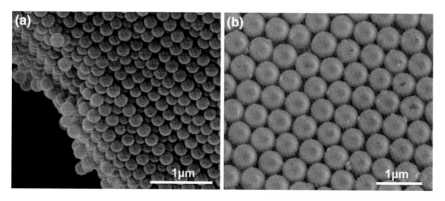

Fig. 2.8 FE-SEM images of **a** multilayers and **b** monolayer of spheres

deposition, dip-coating, and spin-coating, as illustrated in schematic diagrams in Fig. 2.9. The successes of these strategies are dependent on various factors determining quality of the coated monolayer. If the processing conditions are suboptimal, the spheres will simply aggregate to form undesirable clusters, or spread out loosely without any particular order. Due to the low-withdrawal speed in dip-coating and slow evaporation rate in vertical deposition, long processing time (of the order of hours) may be required to establish a monolayer over centimeter-scale areas. Moreover, it is often difficult to maintain precise control of the ambient conditions, such as temperature, humidity, and pressure which strongly affect the evaporation rate. On the other hand, the spin-coating method, which is mainly governed by rotation speed, concentration, size of spheres, and wettability of substrate, is a more reliable method with distinct advantages of higher throughput and better reproducibility. The areas of close-packed monolayer are typically of the order $\sim cm^2$, while the equipment cost of spin-coating technique is low. When these parameters are fine-tuned, monolayer arrays of close-packed nanospheres over centimeters can be obtained within a matter of minutes.

Uniform silica nanospheres are initially diluted in deionized water to produce the optimal volume concentration of ~ 2 %. The diluted colloidal suspension is then mixed with sodium dodecyl sulfate at a volume ratio of 10:1. Introduction of a surfactant lowers the surface tension of the colloidal suspension and thus facilitates the spreading of nanosphere to prevent particle agglomeration or aggregation. The well-mixed colloidal suspension is then carefully dispensed onto the sample surface by mechanical micropipetting or other means. Optimized rotational speed is necessary to balance the centrifugal force with the solvent capillary force. During spin-coating, the excess suspension is gradually flung off and spheres spread laterally, self-assembling into a monolayer of hexagonal-close-packed array across the sample. The coated spheres then act as a lithographic mask and the pillar pattern is transferred to the wafer by inductive-coupled plasma (ICP) etching. The etched

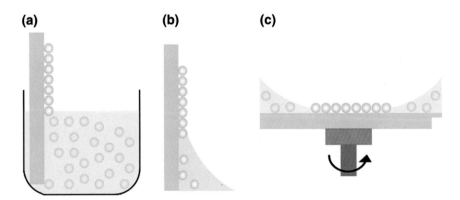

Fig. 2.9 Schematic diagrams showing various nanosphere coating procedures, including **a** vertical deposition, **b** dip-coating, and **c** spin-coating

Fig. 2.10 FE-SEM images of
an HCP nanopillar array
patterned by NSL

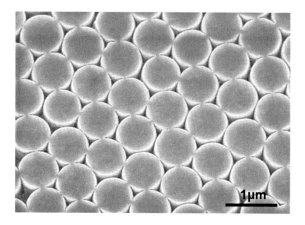

sample is finally immersed in deionized water under sonication to remove the
spheres, leaving behind a nanopillar array. The resultant HCP nanopillar array after
etching is shown in Fig. 2.10.

References

1. Heikkila O, Oksanen J, Tulkki J (2010) The challenge of unity wall plug efficiency: the effects
   of internal heating on the efficiency of light emitting diodes. J Appl Phys 107(3). Artn 033105.
   doi: 10.1063/1.3285431
2. Akasaka T, Gotoh H, Saito T, Makimoto T (2004) High luminescent efficiency of InGaN
   multiple quantum wells grown on InGaN underlying layers. Appl Phys Lett 85(15):3089
   −3091. doi: 10.1063/1.1804607
3. David A, Fujii T, Sharma R, McGroddy K, Nakamura S, DenBaars SP, Hu EL, Weisbuch C,
   Benisty H (2006) Photonic-crystal GaN light-emitting diodes with tailored guided modes
   distribution. Appl Phys Lett 88(6). Artn 061124. doi:10.1063/1.2171475
4. Lester SD, Ponce FA, Craford MG, Steigerwald DA (1995) High dislocation densities in
   high-efficiency gan-based light-emitting-diodes. Appl Phys Lett 66(10):1249−1251. doi: 10.
   1063/1.113252
5. Fujii T, Gao Y, Sharma R, Hu EL, DenBaars SP, Nakamura S (2004) Increase in the extraction
   efficiency of GaN-based light-emitting diodes via surface roughening. Appl Phys Lett 84
   (6):855−857. doi: 10.1063/1.1645992
6. Minsky MS, White M, Hu EL (1996) Room-temperature photoenhanced wet etching of GaN.
   Appl Phys Lett 68(11):1531–1533. doi:10.1063/1.115689
7. Gao Y, Fujii T, Sharma R, Fujito K, Denbaars SP, Nakamura S, Hu EL (2004) Roughening
   hexagonal surface morphology on laser lift-off (LLO) N-face GaN with simple
   photo-enhanced chemical wet etching. Jpn J Appl Phys 243(5A):L637-L639. doi:10.1143/
   Jjap.43.L637
8. Yang CC, Horng RH, Lee CE, Lin WY, Pan KF, Su YY, Wuul DS (2005) Improvement in
   extraction efficiency of GaN-based light-emitting diodes with textured surface layer by natural
   lithography. Jpn J Appl Phys 144(4B):2525–2527. doi:10.1143/Jjap.44.2525
9. Li ZL, Li KH, Choi HW (2010) Mechanism of optical degradation in microstructured InGaN
   light-emitting diodes. J Appl Phys 108(11). Artn 114511. doi:10.1063/1.3517829

10. Jin SX, Li J, Lin JY, Jiang HX (2000) InGaN/GaN quantum well interconnected microdisk light emitting diodes. Appl Phys Lett 77(20):3236–3238. Pii [S0003-6951(00)04546-0]. doi: 10.1063/1.1326479

11. Choi HW, Dawson MD, Edwards PR, Martin RW (2003) High extraction efficiency InGaN micro-ring light-emitting diodes. Appl Phys Lett 83(22):4483–4485. doi:10.1063/1.1630352

12. Choi HW, Chua SJ (2006) Honeycomb GaN micro-light-emitting diodes. J Vac Sci Technol B 24(2):800–802. doi:10.1116/1.2184324

13. Lee YC, Lee CE, Kuo HC, Lu TC, Wang SC (2008) Enhancing the light extraction of (AlxGa1-x)(0.5) In0.5P-based light-emitting diode fabricated via geometric sapphire shaping. IEEE Photonic Tech L 20(5-8):369-371. doi:10.1109/Lpt.2008.916905

14. Lee CE, Lee YC, Kuo HC, Tsai MR, Lu TC, Cwang S (2008) High brightness GaN-based flip-chip light-emitting diodes by adopting geometric sapphire shaping structure. Semicond Sci Tech 23(2). Artn 025015. doi:10.1088/0268-1242/23/2/025015

15. Kao CC, Kuo HC, Huang HW, Chu JT, Peng YC, Hsieh YL, Luo CY, Wang SC, Yu CC, Lin CF (2005) Light-output enhancement in a nitride-based light-emitting diode with 22 degrees undercut sidewalls. IEEE Photonic Tech L 17(1):19–21. doi:10.1109/Lpt.2004. 837480

16. Wang XH, Lai PT, Choi HW (2010) The contribution of sidewall light extraction to efficiencies of polygonal light-emitting diodes shaped with laser micromachining. J Appl Phys 108(2). Artn 023110. doi:10.1063/1.3456445

17. Kim SH, Lee KD, Kim JY, Kwon MK, Park SJ (2007) Fabrication of photonic crystal structures on light emitting diodes by nanoimprint lithography. Nanotechnology 18(5). Artn 055306. doi:10.1088/0957-4484/18/5/055306

18. Oder TN, Shakya J, Lin JY, Jiang HX (2003) III-nitride photonic crystals. Appl Phys Lett 83 (6):1231–1233. doi:10.1063/1.1600839

19. Matsubara H, Yoshimoto S, Saito H, Yue JL, Tanaka Y, Noda S (2008) GaN photonic-crystal surface-emitting laser at blue-violet wavelengths. Science 319(5862):445–447. doi:10.1126/ science.1150413

20. Knight JC (2003) Photonic crystal fibres. Nature 424 (6950):847–851. doi:10.1038/ Nature01940

21. Lai CF, Chi JY, Yen HH, Kuo HC, Chao CH, Hsueh HT, Wang JFT, Huang CY, Yeh WY (2008) Polarized light emission from photonic crystal light-emitting diodes. Appl Phys Lett 92 (24). Artn 243118. doi:10.1063/1.2938885

22. Wierer JJ, Krames MR, Epler JE, Gardner NF, Craford MG, Wendt JR, Simmons JA, Sigalas MM (2004) InGaN/GaN quantum-well heterostructure light-emitting diodes employing photonic crystal structures. Appl Phys Lett 84(19):3885–3887. doi:10.1063/1. 1738934

23. Kim DH, Cho CO, Roh YG, Jeon H, Park YS, Cho J, Im JS, Sone C, Park Y, Choi WJ, Park QH (2005) Enhanced light extraction from GaN-based light-emitting diodes with holographically generated two-dimensional photonic crystal patterns. Appl Phys Lett 87(20). Artn 203508. doi:10.1063/1.2132073

24. Redwing JM, Loeber DAS, Anderson NG, Tischler MA, Flynn JS (1996) An optically pumped GaN-AlGaN vertical cavity surface emitting laser. Appl Phys Lett 69(1):1–3. doi:10. 1063/1.118104

25. Someya T, Werner R, Forchel A, Catalano M, Cingolani R, Arakawa Y (1999) Room temperature lasing at blue wavelengths in gallium nitride microcavities. Science 285 (5435):1905–1906. doi:10.1126/science.285.5435.1905

26. Tamboli AC, Haberer ED, Sharma R, Lee KH, Nakamura S, Hu EL (2007) Room-temperature continuous-wave lasing in GaN/InGaN microdisks. Nat Photonics 1(1):61–64. doi:10.1038/ nphoton.2006.52

27. Mair RA, Zeng KC, Lin JY, Jiang HX, Zhang B, Dai L, Botchkarev A, Kim W, Morkoc H, Khan MA (1998) Optical modes within III-nitride multiple quantum well microdisk cavities. Appl Phys Lett 72(13):1530–1532. doi:10.1063/1.120573

28. Lu TC, Chen SW, Lin LF, Kao TT, Kao CC, Yu P, Kuo HC, Wang SC, Fan SH (2008) GaN-based two-dimensional surface-emitting photonic crystal lasers with AlN/GaN distributed Bragg reflector. Appl Phys Lett 92(1). Artn 011129. doi:10.1063/1.2831716

29. Lai CF, Yu P, Wang TC, Kuo HC, Lu TC, Wang SC, Lee CK (2007) Lasing characteristics of a GaN photonic crystal nanocavity light source. Appl Phys Lett 91(4). Artn 041101. doi:10.1063/1.2759467

30. Oder TN, Kim KH, Lin JY, Jiang HX (2004) III-nitride blue and ultraviolet photonic crystal light emitting diodes. Appl Phys Lett 84(4):466–468. doi:10.1063/1.1644050

31. Motayed A, Davydov AV, Vaudin MD, Levin I, Melngailis J, Mohammad SN (2006) Fabrication of GaN-based nanoscale device structures utilizing focused ion beam induced Pt deposition. J Appl Phys 100(2). Artn 024306. doi:10.1063/1.2215354

32. Lanyon YH, De Marzi G, Watson YE, Quinn AJ, Gleeson JP, Redmond G, Arrigan DWM (2007) Fabrication of nanopore array electrodes by focused ion beam milling. Anal Chem 79 (8):3048–3055. doi:10.1021/Ac061878x

33. Haynes CL, Van Duyne RP (2001) Nanosphere lithography: A versatile nanofabrication tool for studies of size-dependent nanoparticle optics. J Phys Chem B 105(24):5599–5611. doi:10.1021/Jp010657m

34. Ozin GA, Yang SM (2001) The race for the photonic chip: Colloidal crystal assembly in silicon wafers. Adv Funct Mater 11(2):95–104. doi:10.1002/1616-3028(200104)11:2<95::Aid-Adfm95>3.0.Co;2-O

35. Zhang Q, Li KH, Choi HW (2012) Polarized emission from InGaN Light-emitting Diode with Self-assembled opal coating. 2012 12th IEEE conference on nanotechnology (IEEE-Nano)

36. Zhang Q, Li KH, Choi HW (2012) Polarized emission from InGaN light-emitting diodes with self-assembled nanosphere coatings. IEEE Photonic Tech L 24(18):1642–1645. doi:10.1109/Lpt.2012.2211586

Chapter 3
III-Nitride Light-Emitting Diodes with Photonic Crystal Structures

Abstract In this chapter, four works involving the application of self-assembly NSL processes to GaN-based material are introduced, mainly to deal with the light extraction issues of GaN LEDs so as to suppress light confinement through reduction of total internal reflections at the GaN/air interfaces. Two electroluminescence devices incorporating nanosphere-patterned structure are first discussed; the self-assembled array of spheres with varying dimensions serves as a hard mask to form the close-packed PhCs onto the ITO film (Li and Choi, J Appl Phys 110 (5), 2011, [1]) and the intermediate layer of semiconductor (Li et al, Appl Phys Lett 102 (18), 2013, [2]). To extend the function of sphere-pattern array, dimension-adjusting procedure is employed to overcome the restrictions of close-packed patterning and realize photonic bandgap structures. With a well-defined periodic arrangement and with sufficiently large refractive index contrast between GaN and ambient, a wavelength-tunable PBG in the visible region can be opened up, which forbids the propagation of light within a specific range of frequencies and is extremely useful in molding the flow of emitted light from an LED. Two non-close-packed periodic patterns, namely air-spaced (Li and Choi, J Appl Phys 109 (2), 2011, [3]) and clover-shaped (Li et al, Appl Phys Lett 100 (14), 2013, [4]) PBG structures, are highlighted in this chapter.

3.1 InGaN Light-Emitting Diodes with Indium Tin Oxide Photonic Crystal Current-Spreading Layer

Photonic crystal patterns on the indium tin oxide layer of an InGaN/GaN LED are fabricated by NSL in combination with dry etching. The silica spheres acting as an etch mask are self-assembled into HCP monolayer array. After etching, the photonic crystal pattern is formed across the ITO films so that the semiconductor layers are left intact and thus free of etch damages. Despite slight degradation to electrical properties, the ITO-PhC LEDs exhibited enhancements of optical emission power

© Springer-Verlag Berlin Heidelberg 2016
K.H. Li, *Nanostructuring for Nitride Light-Emitting Diodes and Optical Cavities*, Springer Theses, DOI 10.1007/978-3-662-48609-2_3

by as much as 64 % over an as-grown LED. The optical performances and mechanisms of the photonic crystal LEDs are investigated with the aid of rigorous coupled wave analysis and finite-difference time-domain simulations.

3.1.1 Introduction

Without a doubt, significant improvement of light extraction efficiency can be achieved by texturing the surface of LED. However, the choice of substrate for surface patterning is critical issue highly affecting the electrical characterization of devices. When the top p-GaN contact layer involves plasma etching, the plasma damage induced will significantly degrade electrical conduction in the device due to increased ohmic contact resistance and leakage currents, thus sacrificing overall efficiency. Plasma damage of the p-n junction has also been shown to affect device lifetime. Instead of directly processing the GaN layer, surface roughening of the indium tin oxide (ITO) current-spreading layer has been demonstrated as an alternative method for improving light extraction efficiency without degradation of electrical characteristics.

The fabrication and characterization of PhC-on-ITO LEDs patterned by NSL is reported in this section. The self-assembled array of spheres serves as a hard mask for pattern transfer onto the ITO layer, resulting in hexagonally close-packed ITO pillar arrays. Although such PhC structures do not possess a PBG in the visible region, light extraction can be improved via the dispersion effect. The guided modes are effectively diffracted by the periodic refractive lattice. The performance of the ITO-PhC LEDs is evaluated, together with an investigation on the mechanisms involved. The RCWA algorithm and FDTD method are employed to investigate the effect of incorporating PhCs of different dimensions on the performance of LEDs.

3.1.2 Experimental Details

A schematic diagram illustrating the process flow of ITO-PhCs LEDs in this work is depicted in Fig. 3.1. The InGaN/GaN LED wafers are grown on c-plane sapphire substrate by MOCVD, with embedded multiquantum wells designed for emission at around 470 nm. A 200 nm-thick transparent ITO coating is deposited by sputtering as current-spreading layer, as shown in Fig. 3.1a. The ITO-PhC structure is patterned by NSL, beginning with the dispensing of a colloidal suspension onto the surface of the wafer using a micropipette. The colloidal suspension is prepared by mixing silica spheres with mean diameters of 500, 700, and 1000 nm suspended in DI water with an anionic surfactant sodium dodecyl sulfate (SDS). The spheres self-assemble naturally and uniformly across the ITO layer with the aid of spin

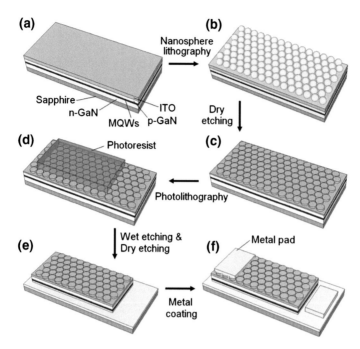

Fig. 3.1 Schematic diagrams showing the process flow; **a** the starting wafer; **b** silica spheres spin coated onto ITO layer; **c** pattern transfer to ITO layer by ICP etching followed by sphere removal; **d** mesa definition by photolithography; **e** exposure of n-GaN region after dry etching; **f** metal pads deposition by e-beam evaporation. Reprinted from Ref. [1] with permission from AIP Publishing LLC

coating at optimized conditions. The rotation speed is varied between 140 to 200 rpm depending on the sphere diameter for duration of 10 min, resulting in the formation of a monolayer of spheres over an area of approximately 8×8 mm^2.

The self-assembled hexagonal close-packed array serves as an etch mask, the pattern of which is subsequently transferred to the ITO layer by ICP etching. The etch parameters are set to 500 W of coil power, 150 W of platen power at 5 mTorr of chamber pressure, using a gas chemistry comprising 15 SCCM of Cl_2 and 10 SCCM of Ar. Photolithographic patterning defines 600 µm by 300 µm mesa regions, followed by dry etching to expose the n-GaN layer. Another photolithographic step is performed to define the contact pad regions for metallization. The p-pads and n-pads are deposited by e-beam evaporation and the wafer is subject to rapid thermal annealing at 500 °C in an N_2 ambient to form ohmic contacts. For comparison, an unpatterned LED of identical dimensions is fabricated alongside. The chips are diced by ultraviolet nanosecond laser micromachining and die bonded onto TO headers, followed by Al wire bonding. The surface morphologies of PhCs LEDs are imaged using field emission scanning electron microscopy.

3.1.3   Surface Morphologies

Figure 3.2a shows an ordered hexagonal monolayer of nanosphere. To minimize the occurrence of defects such as dislocations and vacancies, and to avoid the formation of multiple layers which disrupts the desired hexagonal pattern, the spin coat rotation speed must be optimized. With decreasing sphere dimensions, an increase in spin velocity (this centrifugal force) is required to overcome the viscosity of the suspension. The optimized speeds for achieving high-quality monolayer array are determined to be 200, 160, and 140 rpm for nanosphere diameters of 500, 700, and 1000 nm, respectively. After dry etching, periodic hexagonal close-packed ITO pillar arrays are formed with triangle air gap voids between adjacent pillars exposed, as shown in Fig. 3.2b. The etch depths of pillars are around 100 nm corresponding to an etching duration of 90 s, as estimated from the SEM image in Fig. 3.2c captured at a tilted angle of 30°. Figure 3.2d–f shows the resultant PhC structures after sphere removal with pillar diameters of 500, 700, and 1000 nm, respectively.

3.1.4   Device Characterizations

EL measurements are conducted on the packaged and unencapsulated PhC and unpatterned LEDs by collecting the emitted light with a 2-in. integrating sphere optically coupled to a radiometrically calibrated spectrometer. Figure 3.3 shows plots of EL intensity versus injection current for the devices, from which it is

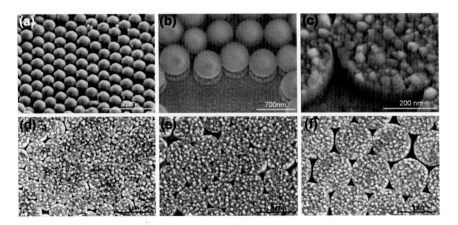

Fig. 3.2 FE-SEM images showing **a** ordered hexagonal monolayer arrays of nanospheres on a GaN LED wafer; **b** the nanosphere-coated sample after ICP etching, **c** close-up view of the ITO nanopillars, and resultant nanopillar arrays with diameters of **d** 500, **e** 700, and **f** 1000 nm. Reprinted from Ref. [1] with permission from AIP Publishing LLC

Fig. 3.3 Light output power
as a function of injection
currents. Reprinted from Ref.
[1] with permission from AIP
Publishing LLC

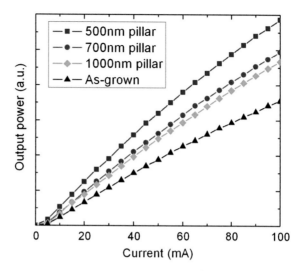

observed that the PhCs LEDs exhibit strong enhancements in light emission over
the unpatterned LED. At an injection current of 100 mA, the output powers of PhCs
LEDs with pillar diameters of 500, 700, and 1000 nm were enhanced by 64.6, 39.1,
and 31.2 %, respectively. The current–voltage (I–V) characteristics of the LEDs are
plotted in Fig. 3.4. The forward voltages at 20 mA dc current are 3.30, 3.28, 3.26,
and 3.25 V, for PhCs LEDs with diameters of 500, 700, and 1000 nm, and the
unpatterned LED, respectively. The slopes of the I–V curves in the linear region
(thus series resistance) are also identical. The I–V data testify to the fact that
nanostructuring of the ITO layer does not degrade electrical characteristics of the
LEDs, an important consideration for minimizing power consumption.

Fig. 3.4 I–V characteristics
of PhCs LEDs and as-grown
LED samples. Reprinted from
Ref. [1] with permission from
AIP Publishing LLC

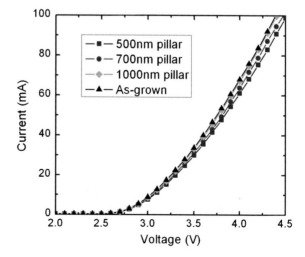

Figure 3.5a–d shows plan view microphotographs of the PhCs LEDs operated at 5 mA, with pillar diameters of 500, 700, and 1000 nm, together with the as-grown LED. As the ITO layer is not degraded by the microstructuring, uniform emission is maintained. The PhCs LEDs also appear brighter with decreasing pillar diameters. Compared with the PhCs LEDs, emission along the edges is significantly stronger than the planar regions from the as-grown LED, since the guided photons are either reabsorbed within the active layer or escape through the sidewalls. The PhCs LEDs, on the other hand, offers enhanced emission intensity over the entire planar surface. To investigate the function of the patterned ITO layer, a reflectivity simulation is performed based on the RCWA algorithm. The defined unit cell is as shown in the inset of Fig. 3.6 and a $\lambda = 450$ nm beam was incident onto the periodic array. At the ITO ($n_{air} = 1.9$)/air ($n_{ITO} = 1.0$) interface, the critical angle determined by

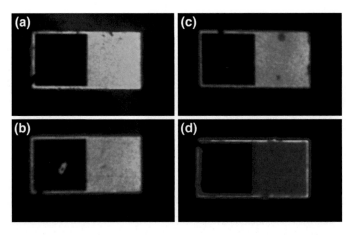

Fig. 3.5 Optical microphotographs showing emission from ITO-PhC LEDs with pillar diameters of **a** 500 nm, **b** 700 nm **c** 1000 nm and **d** the as-grown LED. The devices are biased at 5 mA. Reprinted from Ref. [1] with permission from AIP Publishing LLC

Fig. 3.6 Calculated reflection spectra under varying incident angles. Reprinted from Ref. [1] with permission from AIP Publishing LLC

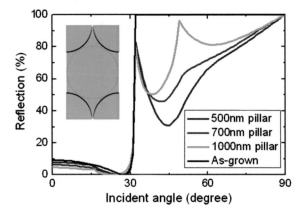

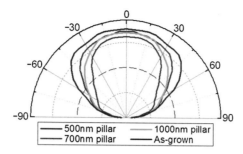

Fig. 3.7 Angular emission patterns of PhCs LEDs and as-grown sample. Reprinted from Ref. [1] with permission from AIP Publishing LLC

$\sin^{-1}(n_{air}/n_{ITO})$ is $\sim 31.8°$ such that incident light rays striking the interface at angles greater than the critical angle are totally reflected, as illustrated in Fig. 3.6. On the other hand, the ITO film incorporating PhCs serves as a light extraction layer for suppression of TIR, so that more photons are capable of escaping from the devices. Angular-resolved emission patterns of the LEDs are determined by collecting EL intensities with a fiber probe at different angles while maintaining a fiber-LED separation is 50 mm. The light collected by the fiber is channeled to an optical spectrometer. The peak EL intensity at each angle is taken to plot the emission pattern, as shown in Fig. 3.7. For the as-grown LED, the intensity drops rapidly beyond $\sim 30°$, while the FWHM divergence angles increase with reducing pillar diameters. The results are consistent with RCWA simulated predictions.

3.1.5 Theoretical Simulations

To further evaluate the effect of PhCs patterns on LEDs, a 3-D FDTD simulation is carried out. Periodic boundary conditions are applied to the x-y plane. The simulated LED structure consists of 200 nm-thick ITO/150 nm-thick p-GaN/40 nm-thick MQWs/2000 nm-GaN. The wavelength of the source was set as 450 nm and the mesh size is 20 nm. A sufficient simulation period was allowed so that light output signal attains steady state. Figure 3.8a shows FDTD simulated emission pattern from an unpatterned LED; photons emitted outside the critical angle are seen to be totally reflected at the flat interface. On the other hand, the incorporation of periodic PhC structures onto the ITO layer is seen to suppress lateral guiding modes and redirect the trapped photons into radiated modes, as illustrated in Fig. 3.8b. The simulated time-resolved light intensity plot in Fig. 3.9 shows that PhC LEDs with pillar diameters of 500, 700, and 1000 nm transmit 61.4, 29.5, and 20.0 % more light than the flattop sample, correlating well with previous experimental and simulated results.

Fig. 3.8 Comparison of
FDTD simulation between the
a as-grown flattop and **b** PhCs
samples. Reprinted from Ref.
[1] with permission from AIP
Publishing LLC

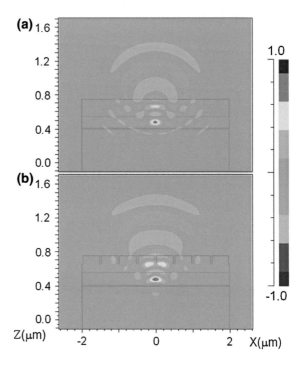

Fig. 3.9 FDTD simulation
results of light output power
of PhCs. Reprinted from Ref.
[1] with permission from AIP
Publishing LLC

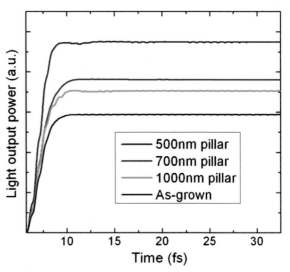

FDTD simulations are also performed to study the effects of varying pillar
heights on the optical output power. The heights of pillars are varied from 0 nm to
160 nm while the other device parameters remain unchanged. The computed output
powers are normalized with respect to that of an unpatterned LED. From the

Fig. 3.10 Plot of FDTD-computed normalized light output power as a function of pillar height for the 500, 700, and 1000 nm PhC LEDs. Reprinted from Ref. [1] with permission from AIP Publishing LLC

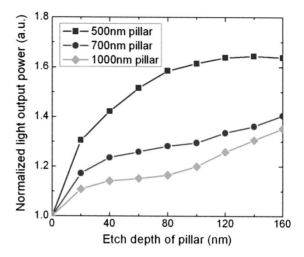

simulated results plotted in Fig. 3.10, it is apparent that taller pillars generally deliver larger optical powers. It is also observed that the rate of increase of output power for the 500 nm PhC slows down significantly after exceeding a height of ∼ 100 nm. As for the 700 nm and 1000 nm PhCs, although gradual increase in output power continues beyond pillar heights of 100 nm, the higher degree of penetration into ITO film would degrade the lateral conductivity of the current-spreading layer and thus the electrical properties. In view of such considerations, the heights of pillar are design to be ∼ 100 nm which is half of the total thickness of the ITO film, in order to maximize overall device performance in terms of both optical and electrical characteristics.

Two-dimensional PhCs are known to promote light extraction in LEDs via two possible mechanisms. If the PhC possesses a PBG along the plane, lateral guiding mode can effectively be eliminated over the range of frequencies covered by the bandgap. However, PhCs comprising close-packed pillar structures as fabricated by NSL do not possess PBGs, as confirmed by the simulated TE and TM band diagrams shown in Fig. 3.11. Of course, it is possible to produce PBG structures using NSL. In this chapter, the fabrication of air-spaced nanopillar structures will be introduced by shrinking the patterned sphere pattern prior to pattern transfer. In this way, a PBG can be induced from such air-spaced pillar structures. In the present study, the "weak" PhCs serve to redirect emission from guided modes into radiative modes [5]. A periodic refractive index is capable of altering the propagation behavior of photon, as described by the dispersion relation $\omega(k)$ with the light line $\omega = k_0 c$ for free space propagation. According to Bloch's theorem, the dispersion curves of Bloch modes are folded at the Brillouin zone boundary, as evident from Fig. 3.11. As a result, the waveguided modes originally located below the light line can be folded to the diffracted mode which are located above the light line, and thus can be extracted, provided the lattice constant is larger than the cutoff (Λ_{cutoff}), which is evaluated by $\Lambda_{cutoff} \approx \lambda/(n_{eff} + 1)$, where n_{eff} is the effective index of the

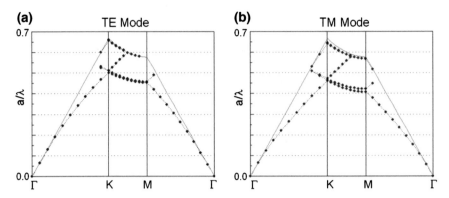

Fig. 3.11 Simulated **a** TE and **b** TE band structures for close-packed pillar arrays. Reprinted from Ref. [1] with permission from AIP Publishing LLC

PhC layer [6]. The lattice constants of the PhCs LEDs described in this study are in the range of 500 nm to 1000 nm, and thus satisfy the cutoff condition.

3.1.6   Conclusion

In summary, we have demonstrated the fabrication of LEDs with a PhC on the ITO current-spreading layer, patterned by NSL. The self-assembled HCP sphere array pattern is transferred to the ITO layer by dry etching. In this way, the semiconductor layers are not degraded in the process. No significant degradation to the ITO layer is observed, so that emission uniformity and good I–V characteristics are maintained. The emission powers increased with decreasing pillar diameters. In particular, the output power of the ITO-PhC LEDs with diameter of 500 nm is enhanced by 64 % compared with the unpattern LED. The significant enhancement can be attributed to dispersion behavior and diffraction property of PhCs. The measured results are verified by simulations based on the RCWA algorithm and the FDTD method.

3.2   III-Nitride Light-Emitting Diode with Embedded Photonic Crystal

A photonic crystal has been embedded within an InGaN/GaN light-emitting diode structure via epitaxial lateral overgrowth of a p-type GaN capping layer. The photonic crystal is a hexagonal close-packed array of nanopillars with various diameters, patterned by nanosphere lithography; the capping layer reconnects the disconnected pillars to form an electron-injection device. The nanopillar approach offers the additional benefit of strain relaxation over conventional airhole structures. Optical properties of the nanostructures and devices are extensively studied through a range of spectroscopy techniques and simulations. The photonic crystal LEDs not only emit more light than regular LEDs, but more importantly, their emission wavelengths are nearly invariant of injection currents, attributed to partial suppression of the built-in piezoelectric in the quantum well regions, representing one grand solution boosting internal and external quantum efficiencies simultaneously.

3.2.1   Introduction

III-nitride LEDs are already more energy efficient than gas-discharge fluorescent lamps, its closest competitor, although not to the extent LEDs are theoretically capable of. Quantum efficiencies, both IQE and LEE, fall short of their potentials, owing to a combination of factors including strain [7, 8] and photon confinement. The piezoelectric field, a consequence of compressively strained InGaN/GaN quantum wells, reduces IQE due to far from optimal overlap between electron–hole wavefunctions [9–11]. On the other hand, the small-angle emission cone due to high refractive index contrast at the semiconductor/air interface gives rise to extraction efficiencies of below 10 %; reabsorption of confined light becomes heat, further jeopardizing IQE. Solutions to such problems are plentiful: GaN substrates [12, 13] offer the perfect solution to lattice-matched epitaxy eliminating strained quantum structures; until such substrates become widely available GaN-on-sapphire remains mainstream technology and the IQE issue remains. On the other hand, structuring of LED chips at the nanoscales promotes light extraction, alleviating EQE constraints. The integration of PhCs has been demonstrated to improve optical performance and directionality control. Although PhCs can be of the form of arrays of airholes or pillars, reported PhC LEDs mostly adopt the former configuration [14], where device processing does not involve planarization. Yet, periodic arrays of pillars are not just equally capable of extracting guided photons [15], but potentially can enhance IQE simultaneously due to the partial strain relaxation. While this sounds attractive, electrical injection to individually isolated nanoscale emitters poses a challenge to device processing. Typically, electrical interconnection of a nonplanar surface involves planarization via gap filling [16] followed chemical mechanical polishing; though feasible, filling the air gaps between adjacent nanopillars reduces the refractive index contrast, disrupting optical properties of the PhC. In this work, an innovative approach of planarization through wafer regrowth is reported. A nanostructured device comprising a pillar-type PhC is planarized through the regrowth of a continuous p-doped contact layer. Such overgrowth techniques produce epitaxial layers with reduced threading dislocation density, ensuring high crystalline quality. With regard to formation of the nanoscale pillar array, a standard NSL process is adopted to defining large-area hexagonal close-packed array, serving as a weak photonic crystal after pattern transfer.

3.2.2 Experimental Details

The starting LED wafer used in this study consists of InGaN/GaN QWs grown by MOCVD on c-plane sapphire substrate. Silica spheres with diameters of 192, 310, and 500 nm are used to assemble hexagonal close-packed monolayers, acting as an etch mask which is subsequently transferred to GaN by dry etching. The NSL process is described in detail in Chap. 2. On top of the nanopatterned structure, a continuous p-doped GaN layer is grown using Epitaxial Lateral Overgrowth (ELO), also by MOCVD. The patterned wafers are thoroughly cleaned in the sequence of acetone, methanol, and deionized water, prior to being loaded into the MOCVD chamber. The 300 nm-thick p-type GaN ELO layer is grown at a temperature of 1070 °C at a pressure of 60 Torr, encapsulating the PhC structure. A mixture of N_2 and H_2 are used as carrier gas, while Trimethylgallium (TMGa), Cp_2Mg, and NH_3 are the Ga, dopant, and N sources, respectively. The growth temperature is subsequently decreased to 800 °C whilst maintaining a N_2 ambient for 30 min to activate the Mg dopants. A Ni/Au (10 nm/10 nm) current-spreading layer is deposited over the p-GaN ELO layer by electron beam evaporation, followed by contact alloying by annealing at 600 °C in oxygen ambient.

The LED emissive regions with areas of $400 \times 200 \ \mu m^2$ are defined by photolithography and dry etched to expose the n-GaN layer. Another photolithographic process defines the p-pad and n-pad regions for contact metallization. A bilayer of Ti/Au (40/200 nm) is e-beam evaporated, followed by thermal annealing at 550 °C in N_2 for 5 min to form ohmic contacts. The chips diced by ultraviolet nanosecond laser micromachining are then mounted and wire bonded onto TO headers. The surface morphologies of the devices are investigated by field emission scanning electron microscope. The samples are excited using a DPSS 349 nm laser operating at a repetition rate of 1 kHz and pulse duration of 4 ns. The PL signals are dispersed by an Acton SP2500A 500 mm spectrograph and detected by a PI-PIXIS open-electrode charge-coupled device (CCD).

The schematic diagram in Fig. 3.12 sketches the skeletal structure of the LED with embedded PhCs. Planar and cross-sectional views of the embedded nanostructure are captured using FE-SEM and shown in Fig. 3.13a–b, clearly reveal successful planarization of the nanopillars via the ELO layer. The electrical, optical, and structural properties of the packaged device are comprehensively characterized, complemented by FDTD simulations for the optical properties of the PhC structure and finite element strain simulations, with the objective of demonstrating how the strain-relaxed pillars play a role in enhancing performance of the LED.

3.2.3 Optical Properties of Photonic Crystal

The enhancement ratios of the integrated PL signal for the PhCs before and after ELO, compared with the as-grown, are plotted in Fig. 3.14a. The PL intensity is

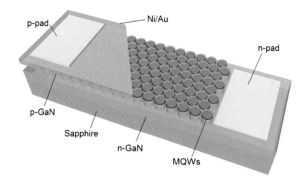

Fig. 3.12 Schematic diagram illustrating the resultant LED with embedded PhCs. Reprinted from Ref. [2] with permission from AIP Publishing LLC

Fig. 3.13 FE-SEM images showing **a, b** the regrowth GaN layers covering the pillar arrays; **c** the cross-sectional view of regrown device. Reprinted from Ref. [2] with permission from AIP Publishing LLC

raised by as much as ~ 1.9 for $d = 192$ nm; the factor decreases with increasing d. Such close-packed pillar arrays do not possess photonic bandgaps in the visible region; instead they operate as 'weak' PhCs due to folding of dispersion curves of Bloch modes at the Brillouin zone boundaries. The enhancement effects are weakened with insertion of the ELO layer, attributed to absorption. The PL spectra of the PhCs after regrowth, plotted in Fig. 3.14b, reveal spectral blueshifts as d decreases, tentatively attributed to strain relaxation.

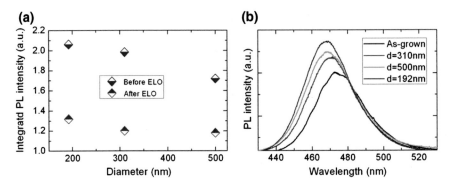

Fig. 3.14 **a** Integrated PL intensity of samples with different diameters before and after ELO. **b** PL spectra of nanopillar arrays. Reprinted from Ref. [2] with permission from AIP Publishing LLC

The optical characteristics of the PhC with and without the ELO layer are further studied through 3-D FDTD simulations. The QWs are modeled as light sources inserted within each pillar, as illustrated in the diagrams depicted in Fig. 3.15a, b. In the simulation, the propagation of waves from the sources is detected and analyzed with a wide-field planar monitor placed on the top of the device models. The simulated data is plotted in Fig. 3.15c, also summarized in its inset. Consistent with the measured PL data, insertion of the ELO layer does reduce light extraction

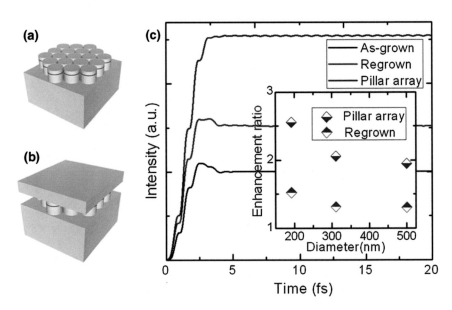

Fig. 3.15 **a** Pillar array and **b** regrown models for 3-D calculation. **c** Simulation results showing the output intensity versus time. Reprinted from Ref. [2] with permission from AIP Publishing LLC

Fig. 3.16 Simulation results of the imperfect PhCs with different size deviations. Reprinted from Ref. [2] with permission from AIP Publishing LLC

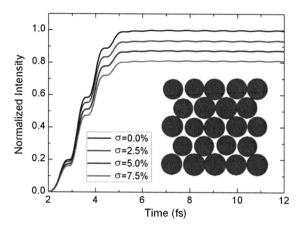

probabilities but light output remains above as-grown levels. It is also noted that the simulated enhancement factors are, on average, 15 % higher than the measured values.

The transmission characteristics of the PhC may be diminished by the presence of defects within the pillar array, a common feature of self-assembly techniques including NSL. Dimensional nonuniformities amongst spheres give rise to point defects as illustrated in the inset of Fig. 3.13. The role of imperfections in the PhC on its optical properties is investigated through additional simulations, conducted using modified models containing pillars with diameter deviations (σ) of ± 2.5, ± 5, and ± 7.5 % from the nominal value of 192 nm. The simulated results, as shown in Fig. 3.16, show that imperfect PhCs extract as much as 20 % less light compared to perfect arrays. Since the diameters of the nanospheres used in this work are known to have deviations of up to ± 7.5 %, the observation of lower-than-predicted PL intensities is justified.

3.2.4 Raman Spectroscopy

While light output may not be maximized, the embedded PhC structure benefits from strain relaxation, a major advantage of this proposed design over airhole-type PhCs. The lattice mismatch between GaN and sapphire substrate, together with yet another lattice mismatch between the well and barrier layers, induces strain in the QWs which significantly reduces internal quantum efficiencies. Nevertheless, strain relaxation occurs at surface boundaries as the surface atoms are less constrained by surrounding material. In nanostructures such as a nanopillar, the surface area to volume ratio is enlarged, so that the extent of strain relaxation becomes proportionally large too. Micro-Raman spectroscopy is employed to detect strain states in the nanopillar PhCs. The 325 nm line of a He-Cd laser, with shallow penetration in GaN, is chosen for Raman excitation so that the signals originate from the near-surface region. The E_2 (TO) phonon peaks from the 500, 310, and 192 nm ELO-PhC and as-grown samples extracted from the Raman spectra in Fig. 3.17a are centered at 569.01, 568.48, 568.28, and 570.39 cm^{-1}, respectively; the redshifts indicate some extent of strain relaxation. The amount of stress can be estimated by $\Delta\omega_{E2} = K_R\sigma$, where σ is the in-plane biaxial stress, K_R a proportionality factor, and $\Delta\omega_{E2}$ is the shift of the E_2 (TO) phonon peak with respect to strain-free GaN, obtained as 567.5 cm^{-1} from free-standing GaN. Using K_R of 4.2 cm^{-1}GPa^{-1}, the compressive stresses are evaluated as 0.360, 0.233, and 0.186 GPa for pillar diameters of 500, 310, and 192 nm, respectively. As predicted, pillars of smaller diameters exhibit larger degrees of strain relaxation.

Raman measurements are also conducted on an identical set of nanopillar structures without the ELO cap layer; the results are summarized in Fig. 3.17b. A slight reduction of redshift in a range of 0.3–00.16 cm^{-1} is observed from samples after ELO regrowth, compared to nanopillar structures without the cap layer. Is that inference of increased strain in the pillars after regrowth? Absolutely not, as the Raman signal comes mainly from the 300 nm p-GaN cap layer. As a

Fig. 3.17 a Micro-Raman spectra for the regrown (ELO) and as-grown samples; **b** The Raman frequency as a function of pillar diameter. Reprinted from Ref. [2] with permission from AIP Publishing LLC

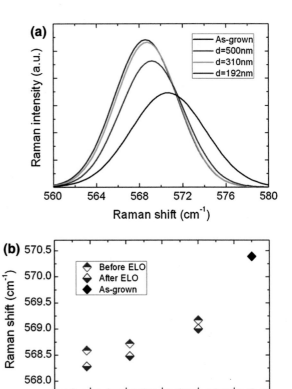

matter of fact, the results indicate that the p-GaN layer is lightly strained (much less than the as-grown), a distinctive feature of ELO layers. This is not surprising as ELO is originally developed for the growth of high-quality, low-dislocation GaN epilayers.

3.2.5 Device Characterizations

With the growth of a cap layer bridging isolated pillars it becomes possible to inject currents into devices embedding PhCs. Figure 3.18 plots electroluminescent (EL) output power of the LEDs as a function of injection current. The PhC LEDs with d = 192, 310, and 500 nm emit 26.63, 17.46, and 11.74 % more light than the reference unpatterned LED. Since the majority of photons emitted from the MQWs are trapped within the device due to total internal reflection, the trapped photons would either be reabsorbed by active region or escape from the sidewalls. On the other hand, the embedded PhCs alter the propagating direction of photons. The improvement of light extraction is attributed to the diffraction and scattering caused by the periodic change in the refractive index.

The incorporation of nanostructures very often degrades electrical properties of the devices. From the I–V plots in Fig. 3.19, the leakage currents measured at a reverse bias of 5 V are 51.2, 45.4, and 21.9 µA for the PhC LEDs of increasing d, being an order of magnitude higher than that of the as-grown LED (2.6 µA), commensurate with sidewall surface areas which act as leakage current pathways. Figure 3.20 shows a plan view of seven adjacent pillars; the total surface areas of the planar and sidewalls of pillars in the blue-shaded region are $3\pi d^2/4$ and $3\pi dH$,

Fig. 3.18 Light output power as a function of injection currents. Reprinted from Ref. [2] with permission from AIP Publishing LLC

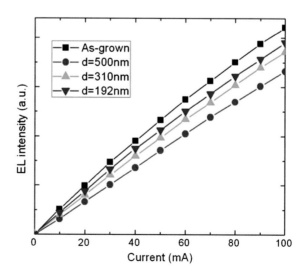

Fig. 3.19 I–V characteristics of resultant devices. Reprinted from Ref. [2] with permission from AIP Publishing LLC

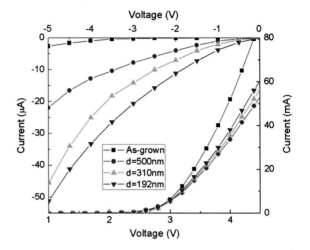

Fig. 3.20 shows a plan view of schematic diagram of 7 adjacent pillars. Reprinted from Ref. [2] with permission from AIP Publishing LLC

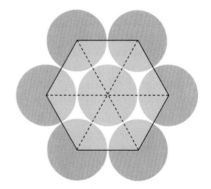

respectively, where H is the height, giving an aspect ratio of *4H/d*, on which the magnitude of leakage currents would depend. The aspect ratios for $d = 192$, 310, and 500 nm are calculated to be 2.60 : 1.61 : 1, which correlate with the ratios of leakage currents according to diameters.

The emission center wavelength of an InGaN LED will not only depend on the bandgap energy of the well layer but is also strongly affected by the large piezo-electric field arising from strain in InGaN/GaN QWs. The triangular-shaped potential well caused by the piezoelectric field and spontaneous polarization tilts the band alignment separating the electron and hole wavefunctions, causing spectral redshifts and reduction of radiative recombination rates. To singulate dimensional effects on the emission spectrum, the LEDs are driven in pulsed mode (1 µs, 1 kHz) to minimize self-heating effects. The emission spectra of the as-grown LED in Fig. 3.21 (left) indicate significant blueshift of 13.31 nm (74 meV) as the bias currents increase from 10 to 115 mA. With increasing currents, the injected carriers

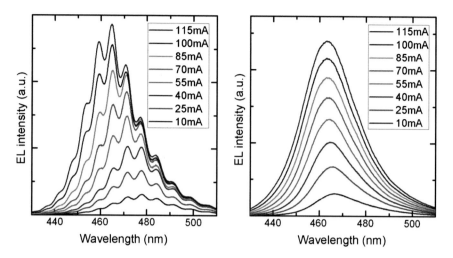

Fig. 3.21 Room temperature electroluminescence spectra of as-grown (*right*) and regrown LEDs with d = 192 nm (*left*) operating at varying injection current. Reprinted from Ref. [2] with permission from AIP Publishing LLC

partially screen the polarization field; such screening reshapes the potential function back to a rectangular profile, leading to the observed spectral blueshift. The filling of localized states in the well and barrier layers may yet be another factor leading to the reduction of effective carrier separation. However, a gradual saturation of localized states would lead to a reduction in quantum efficiency caused by enhanced capture by nonradiative recombination centers.

The L-I data in Fig. 3.17 indicate that the LEDs maintain gradual increases in light emission at high injection currents. In view of this observation, the spectral blueshift is mainly attributed to the screening of the polarization field, increasing electron–hole wavefunction overlap and consequently quantum efficiency. The emission spectra for the d = 192 nm PhC LED is presented in Fig. 3.21 (right); the peak wavelength is nearly invariant of driving currents, signifying that the LED does not suffer from strain effects. The spectra is also characterized by the lack of interference fringes as compared to the as-grown, indicating Fabry–Pérot oscillations have been effectively suppressed by the embedded PhC.

A plot of peak wavelength with respect to bias currents for the LEDs is shown in Fig. 3.22. When d shrinks from 500 nm to 192 nm, the extent of spectral blueshift reduces from 8.88 nm to 4.67 nm over the current range of 10 to 115 mA, compared to 13.31 nm spectral shift of the as-grown LED over the same current range. Such observations of diminishing blueshifts indicate that nanostructuring has indeed partially relaxed the strain and weaken the piezoelectric field, leading to band gap renormalization, the degree of which become increasingly pronounced with decreasing dimensions.

Fig. 3.22 The position of peak wavelength as a function of injection current. Reprinted from Ref. [2] with permission from AIP Publishing LLC

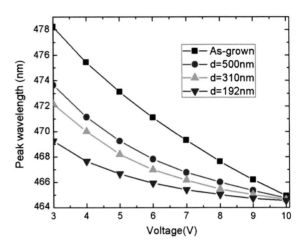

3.2.6   Strain Analysis

To understand the influence of strain on emission characteristics of InGaN QWs, the spectral shift phenomenon is studied by calculating the biaxial in-plane strain and strain-induced piezoelectric polarizations. The biaxial strain (ε_{xx}, ε_{yy}) and strain along the c-axis (ε_{zz}) in a wurtzite structure are generally expressed as [11]

$$\varepsilon_{xx} = \varepsilon_{yy} = \frac{a_s - a_e}{a_e}, \tag{3.1}$$

and

$$\varepsilon_{zz} = -\frac{2C_{13}}{C_{33}} \times \varepsilon_{xx} \tag{3.2}$$

where a_s and a_e are the lattice constants of the GaN barrier and InGaN well layers, while C_{13} (100.8 GPa) and C_{33} (368.8 GPa) are the stiffness constants. By assuming the GaN barrier to be fully relaxed the in-plane strain of the InGaN layer is evaluated as −2.2 %; the negative sign indicates that the layer is under compressive strain. By making use of the calculated values using Eqs. (3.1) and (3.2), the shift of bandgap energy can be estimated by

$$\Delta E = (a_z - D_1 - D_3)\varepsilon_{zz} + (a_x - D_2 - D_4)(\varepsilon_{xx} + \varepsilon_{yy}), \tag{3.3}$$

where the deformation potentials for InGaN are $a_z - D_1 = -5.698$, $a_x - D_2 = -9.03$, $D_3 = 5.592$, $D_4 = -2.79$. To obtain more accurate results, these parameters are determined by linear interpolation between GaN and InN using Vegard's law. The determined relation between ε_{xx} and emission peak is indicated

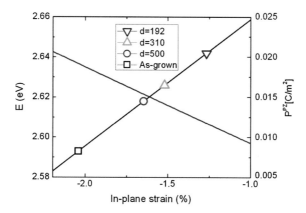

Fig. 3.23 Energy change in the emission peak as a function of biaxial in-plane strain (*black*); and piezoelectric polarization as a function of in-plane strain, using the analytic expressions (*blue*)

by the black line in Fig. 3.23. Based on the measured emission peaks, the corresponding strains for d = 192, 310, and 500 nm are −1.266, −1.519, and −1.646 %, respectively; a reduction of pillar dimensions is accompanied by weakening of the in-plane strain. The PL peak can shift by as much as 0.139 eV if the strain is totally eliminated ($\varepsilon_{xx} = 0$). Moreover, it is noteworthy that the piezoelectric polarization can be reduced by relaxation of compressively biaxial stress, as directly described by equation [17]

$$P_z = e_{31}\varepsilon_{xx} + e_{31}\varepsilon_{yy} + e_{33}\varepsilon_{zz}, \tag{3.4}$$

Since the linear interpolation may underestimate the piezoelectric polarization, especially for alloys with high indium concentrations, the calculated nonlinear piezoelectricity of the binary compounds can be described by the relations [18] (in C/m^2):

$$\begin{aligned} P^{PZ}_{InGaN} &= xP^{PZ}_{InN} + (1-x)P^{PZ}_{GaN}, \\ P^{PZ}_{GaN} &= -0.775\varepsilon + 10.37\varepsilon^2, \\ P^{PZ}_{InN} &= -1.477\varepsilon + 6.837\varepsilon^2, \end{aligned} \tag{3.5}$$

The blue line in Fig. 3.23 shows the piezoelectric polarization of InGaN grown on relaxed GaN buffer layer versus in-plane strain as calculated by Eq. (3.5), which indicates that smaller pillar dimensions give rise to lower compressive strain, further weakening the strain-induced piezoelectric polarizations. A more pronounced reduction of the separation of electron and hole wavefunctions would lead to higher emission energy, explaining the observed blueshifting trends.

A 3-D simulation has also been performed employing finite element approach to map strain distributions along the QW layers, based on a model containing four pairs of 192 nm-disk-shaped QWs. Cross-sectional view of strain distribution along

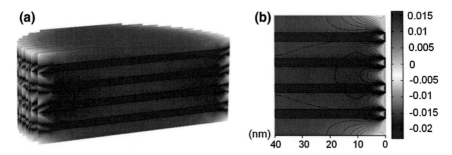

Fig. 3.24 a The cross-sectional planes simulated from the 3-D model. **b** The enlarged plane showing strain distribution at the disk edge. Reprinted from Ref. [2] with permission from AIP Publishing LLC

the various layers is depicted in Fig. 3.24a, revealing that a higher degree of relaxation occurs toward the disk edges while the central region remains strongly strained. Similar degrees of relaxation occur at the edges independent of pillar diameters. Figure 3.24b shows a high-resolution close-up strain map of the sidewall region; the extent of compressive strain gradually increases over a distance of ~ 20 nm from the edge before reaching its maximum value. Consequently, the observed blueshift of spectral peaks suggests that the majority of EL signals originate from the strain-relaxed region surrounding the nanopillars.

3.2.7   Conclusion

In summary, an InGaN LED with an embedded PhC is demonstrated. The PhC is fabricated by etching through an NSL pattern, followed by ELO regrowth. The optical performances of the PhC LEDs have been studied and compared with FDTD simulation results. The periodic ordered nanopillar structure not only promotes light extraction, but also partially suppresses the piezoelectric field through strain relaxation of the InGaN/GaN QWs. Micro-Raman spectroscopy data affirms strain relaxation in the nanopillar samples. The influence of strain on emission wavelength, as well as biaxial in-plane strain mapping has also been investigated with the aid of simulations utilizing the finite element method.

3.3   Air-Spaced GaN Nanopillar Photonic Bandgap Structures

In this section, the fabrication of ordered hexagonal arrays of air-spaced GaN nanopillars is reported. A self-assembled two-dimensional silica nanosphere mask was initially formed by spin coating. Prior to pattern transfer to the GaN substrate, a silica-selective dry etch recipe was employed to reduce the dimensions of the nanospheres, without shifting their equilibrium positions. This process step was crucial to be formation of air-spaced hexagonal arrays of nanospheres, as opposed to close-packed arrays normally achieved by NSL. This pattern is then transferred to the wafer to form air-spaced nanopillars. By introducing air gaps between pillars, a photonic bandgap in the visible region can be opened up, which is usually nonexistent in close-packed nanopillar arrays. The photonic bandgap structures were designed using the Plane Wave Expansion algorithm for bandstructure computations. The existence and positions of bandgaps have been verified through optical transmittance spectroscopy, which correlated well with predictions from simulations. From photoluminescence spectroscopy, a fourfold increase in PL intensity was observed compared to an as-grown sample, demonstrating the effectiveness of well-designed self-assembled PBG structures for suppressing undesired optical guiding mode via photonic bandgap and for promoting light extraction. The effects of defects in the nanopillar array on the optical properties are also critically assessed.

3.3.1 Introduction

Conventionally, PhC are patterned as ordered arrays of recessed air holes or pro-truding pillars. NSL is a practical alternative approach toward large-scale nanofabrication and with the capability of forming 2-D and 3-D PhCs. Uniform spheres are capable of self-assembling into hexagonal arrays over large areas but the ability of spheres to spontaneously form close-packed structures acts also as its limitation: PhCs require alternating layers of different materials with defined and constant separations, and a close-packed structure is obviously not favorable for achieving this. Establishing finite spacing between individual spheres is a critical step toward realizing PBG structures. Formation of PBG structures inhibits all wave vectors within the PBG [19] and thus promotes light extraction by diffracting waveguided modes out with the semiconductor, as illustrated in Fig. 3.25. An air-spaced nanopillar PhC structure with a wavelength-tunable PBG is proposed. Prior to pattern transfer to the wafer, silica nanospheres are shrunk by a selective dry etch process. As a result, spacing is induced between spheres on the plane without altering the sites of spheres; therefore the packing of the modified array remain largely regular. This dimension-adjusting procedure overcomes the restrictions of close-packed patterning to achieve low-cost, high-efficiency, and PBG-tunable nanopillar arrays, which has been applied to InGaN LED for realizing enhancement of light extraction.

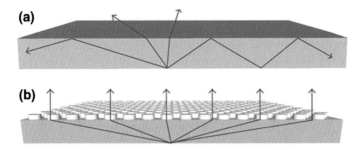

Fig. 3.25 **a** A large fraction of the light is trapped within the semiconductor layer due to total internal reflections. **b** PBG structures promote light extraction by diffracting waveguided modes out of the layer. Reprinted from Ref. [3] with permission from AIP Publishing LLC

3.3.2   Experimental Details

Figure 3.26 illustrates the proposed fabrication process of air-spaced nanopillar arrays by NSL. The hybrid NSL process was performed on III-nitride LED wafers consisting of InGaN/GaN MQWs grown by MOCVD on c-plane sapphire substrate. Uniform silica nanospheres, suspended in deionized water with mean diameters of 192 nm, were mixed with SDS at a volume ratio of 10 : 1. SDS acted as a surfactant to reduce the surface tension of water and thus facilitated the spreading of particles to prevent clustering. 1.5 μL of well-mixed colloidal suspension was then dispensed onto a sample surface by mechanical micropipetting. The nanopheres spread laterally upon spin coating at 1000 r.p.m. for 5 min, self-assembling into a monolayer hexagonal close-packed array across the sample. At this time, the pattern can be

Fig. 3.26 Schematic diagrams illustrating the fabrication flow. **a** Silica nanospheres are coated onto the surface of a GaN wafer, forming close-packed arrays. **b** Shrinkage of spheres using RIE. **c** Pattern transfer to GaN using ICP etching. **d** Silica residue removal

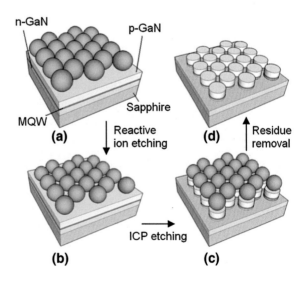

(a)   **(b)**

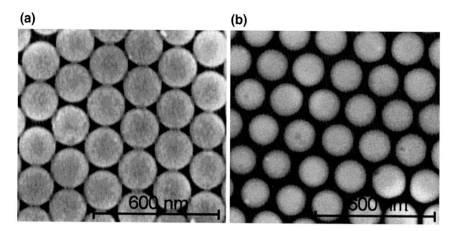

Fig. 3.27 FE-SEM images showing **a** the original ordered close-packed sphere array and **b** the air-spaced sphere array

transferred to the LED wafer to form close-packed nanopillars, as demonstrated from the previous sections. In this work, the nanosphere-coated wafer was subject to reactive ion etching using CHF_3-based plasmas at low RF power prior to pattern transfer. This choice of etchant gas ensured that the silica spheres are selectively etched, without affecting the GaN substrate. During the etching process, the dimensions of the silica nanospheres are reduced. The RF power was maintained low in order to avoid overheating (causing distortion of sphere geometry), and translation of spheres (destroying orderliness of packing). Due to dimensional reduction, an air gap is induced between spheres, as illustrated in Fig. 3.27. The shrunk nanospheres then served as an etch mask and the pattern was subsequently transferred to the GaN wafer by ICP etching using Cl_2 chemistry to form an air-spaced nitride nanopillar array. The ICP and platen powers were maintained at 300 W and 100 W while the chamber pressure was fixed at 5 mTorr. The etch depth was approximately 350 nm after etching for 75 s. The spacing between nanopillars was exposed as imaged by FE-SEM in Fig. 3.28. The nanospheres were subsequently removed by sonification in DI water, leaving behind the nitride air-spaced nanopillar array.

Optical transmittance measurements were conducted to verify the existence and position of a PBG. The incident beam from a high power broadband solid-state plasma light source (Thorlabs HPLS-30-03) was collected by an optical fiber and focused onto the samples in planar direction. The transmitted beam was collected and channeled to an optical spectrometer via another fiber. The optical properties of the PhC structures were further evaluated by time-integrated PL at room temperature. A DPSS UV laser at 349 nm was used as an excitation source (120 µ, 1 kHz) while the PL signal was coupled to a spectrometer comprising an Acton SP2500A 500 mm spectrograph and a Princeton Instrument PIXIS open-electrode CCD via an optical fiber bundle, which offers optical resolutions of better than 0.1 nm.

Fig. 3.28 FE-SEM images
showing the air-spaced
nanopillar array after ICP
etching

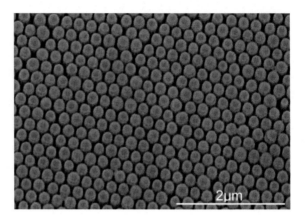

3.3.3 Designing the PhC Structure

To design and predict the existence of a PBG in a nitride nanopillar array, band
diagrams were computed using Rsoft BandSOLVE, which employs the Plane Wave
Expansion algorithm for band computations. The supercell technique was utilized
during three-dimensional PWE simulations. We begin with the simulation of a
close-packed nanopillar array, the structure obtained naturally with NSL. In this
case, the periodicity of the array is considered to be the diameter of a single
nanosphere. The positions of the PBG (a/λ) for the Transverse Electric (TE) modes
were computed as a function of sphere diameter (between 100 to 900 nm) and
plotted in Fig. 3.29, since emissions from InGaN/GaN MQWs are dominated by TE
modes [20]. The plot shows that the TE-PBGs are mainly located in the ultraviolet
and infrared regions of the spectrum. For visible InGaN/GaN LED applications, the
PBG should obviously be located within the visible spectrum to achieve any

Fig. 3.29 Plot of simulated
TE photonic bandgap as a
function of diameters of
close-packed spheres.
Reprinted from Ref. [3] with
permission from AIP
Publishing LLC

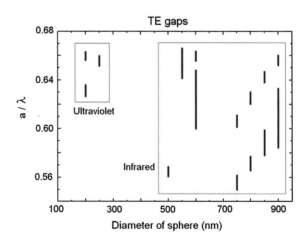

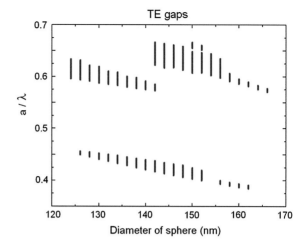

Fig. 3.30 Plot of simulated TE photonic bandgap as a function of diameter of sphere (with a pitch of 192 nm). Reprinted from Ref. [3] with permission from AIP Publishing LLC

beneficial effects. Of course, even in the absence of a PBG, the close-packed nanopillar structure can still be used to increase light extraction via dispersive and geometrical effects [21], albeit with reduced effectiveness.

In spite of the said limitations, a PBG can still be introduced into the visible spectral region by modifying the NSL process to enlarge the physical gaps between nanopillars to produced air-spaced nanopillar arrays, as described in the section on experimental details. To demonstrate this concept, our experiments and computations were carried out using nanospheres with initial nominal diameters of 192 nm with variations of ±2 nm. The positions of the TE bandgaps for shrunk nanopillar arrays of diameters 120–170 nm, with a fixed pitch of 192 nm, were computed and shown in Fig. 3.30. By reducing the diameters of spheres whilst maintaining their pitch, the positions of the TE-PBG shift accordingly. As we are interested in the visible region, the range of frequencies (a/λ) of interest lie between 0.39 and 0.45, corresponding to wavelengths of between 426 to 492 nm. From the simulation results in Fig. 3.30, it can be deduced that the diameters of spheres should be reduced into the range of 126–162 nm, coinciding with the range of InGaN/GaN MQWs emission centered at 455 nm for the material used in our experiments.

3.3.4   Fabrication and Optical Characterization of the PhC Structure

The proposed air-spaced nanopillar structure was fabricated using a modified NSL process. The self-assembled spheres, which acted as a sacrificial masking layer, were etched for dimensional reduction. During this first etching step, the silica spheres are etched without pattern transfer to the wafer. As the dry etch process is directional, the rate of etching in the vertical direction is faster than in the lateral plane. Gradually, the incident ions trim the diameters of the spheres, opening up air gaps between spheres. However, under prolonged etching, the geometry of spheres may become irregular in shape and adjacent spheres tend to aggregate, resulting in disruption of order in the arrays, due to physical bombardment of ions and accumulation of heat. Figure 3.31a, b shows FE-SEM images of nanosphere arrays after etching at 300 W and 70 W of RF powers, respectively. To produce nanopillar arrays of desired order and uniformity, moderate etch conditions are required to minimize defect formation. Based on our calculations, four sets of samples were developed with incremental etch durations. After pattern transfer to the wafer and residue removal, four samples with air-spaced nanopillar structures were produced. The FE-SEM images in Fig. 3.32(a) i–iv illustrate nanopillar arrays etched for durations of 14, 12, 10, and 8 min, respectively; their diameters are roughly equal to 130, 140, 150, and 160 nm, respectively.

Plan view microphotographs in Fig. 3.32(b) i–iv illustrate physical color changes observed from the sample surface in the normal direction, changing from purplish blue to greenish blue with decreasing pillar diameter. Light spots are attributed to larger area of defects which are unpatterned. To verify the existence and position of the PBG, an optical transmission measurement in the planar direction was conducted. Figure 3.32(c) i–iv shows the measured transmission spectra. Four distinct transmission minimas were observed at the wavelengths of 490.59, 471.49, 444.51, and 431.96 nm, respectively, with decreasing sphere diameters of 160, 150, 140, 130 nm. As the PBG structure forbids lateral propagation at the range of wavelengths within the bandgap, propagation of the incident beam along the plane is restricted, giving rise to reduced transmission at those wavelengths. For light with wavelengths beyond the bandgap region, laterally propagating photons do not

(a)   **(b)**

Fig. 3.31 FE-SEM images showing nanosphere arrays after dry etching at **a** 300 W and **b** 70 W RF powers. Reprinted from Ref. [3] with permission from AIP Publishing LLC

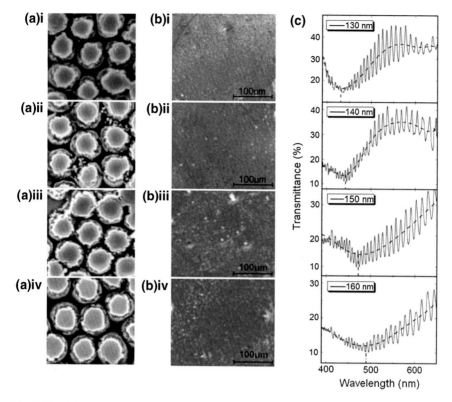

Fig. 3.32 a FE-SEM images showing air-spaced nanopillar arrays that have been RIE etched for durations of *i* 14, *ii* 12, *iii* 10, and *iv* 8 min; **b** Optical microphotographs of nanopillar arrays with diameters of *i* 130 nm, *ii* 140 nm, *iii* 150 nm, *iv* 160 nm, and **c** measured transmission spectra from the respective nanopillar arrays. Reprinted from Ref. [3] with permission from AIP Publishing LLC

experience PBG confinement effects. The transmission spectra, which indicate the PBG positions, correlated well with the PWE stimulated results in Fig. 3.30.

3.3.5 Enhancement of PL Intensity in PhC Structures

The optical effects of incorporating these PhC structures onto LED wafers were evaluated by PL measurement. Figure 3.33 shows measured PL spectra from the four nanopillar arrays, together with the PL spectra from an as-grown sample. Interference fringes in the spectrum of the as-grown sample indicate Fabry–Pérot modes due to vertical optical confinement. Majority of the photons generated by the MQWs remain trapped within the wafer [22], forming standing waves which are subsequently reabsorbed. Such oscillations were clearly suppressed in the

Fig. 3.33 PL spectra from air-spaced nanopillar arrays, compared with an as-grown sample

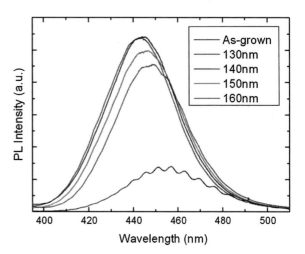

nanopillar samples with diameters of 130, 140, 150, and 160 nm due to diffraction of the guided modes by the PhC, together with the observation of significant PL intensity enhancements, demonstrating that PhCs can indeed play a remarkable role in manipulating spontaneous emission by suppressing unwanted optical modes via the PBG. Compared to the as-grown sample with peak emission wavelength at 455 nm, the peak emission wavelengths from the nanopillar samples were centered at 443.79, 444.05, 446.84, and 448.46 nm, a systematic blueshift with respect to their diameters, which can be attributed to the position of the PBG. The spectral shifts were consequential of the overlap between the MQW emission band and the PBG. With decreasing pillar diameters, the position of the PBG was shifted from 490.59 nm to 431.96 nm. As a result, the spectra contents in the shorter wavelength region were enhanced to a greater extent, giving rise to an apparent spectral blueshift. For the 140-nm-diameter nanopillar array, a fourfold increase in PL intensity was observed since the position of the PBG coincided with the emission wavelength, as evidenced through the computed TE and TM bandstructures in Fig. 3.34. The TE bandgap (the TE mode dominates in InGaN/GaN MQW LEDs) occurs in the frequency range between a/λ of 0.4211 to 0.4402 within the light line of air, corresponding to wavelengths between 436.17 nm and 455.95 nm. It indicates the PBG band corresponding to the 140 nm nanopillar array overlaps optimally with the MQW emission band, correlating well with PL measurements.

3.3.6   The Effects of Disordering in Nanopillar Arrays

With the technique of NSL, the formation of close-packed hexagonal arrays relies on the packing and dimensions of self-assembling microspheres. To achieve a close-packed monolayer array by spin coating, the sphere diffusion rate and

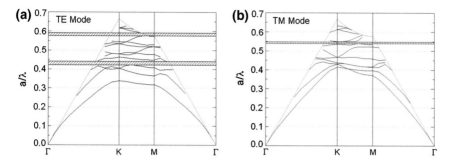

Fig. 3.34 Computed **a** TE and **b** TM band structures for an air-spaced nanopillar array with pillar diameter and pitch of 140 nm and 192 nm, respectively. Reprinted from Ref. [3] with permission from AIP Publishing LLC

concentration of the suspension play important roles in determining the coverage area. The former factor involves a balance between centrifugal forces controlled by rotation speed and surface tension forces which can be reduced by adding the surfactant SDS. Nevertheless, the presence of defects including point defects, line defects, and nonuniformity of sphere diameters are inevitable. Point and line defects are naturally and randomly formed during the self-assembly process. These defects are then transferred to the GaN wafer during etching, affecting the optical properties of the PhC, which is investigated and reported in this section.

The nanospheres were shrunk with a CHF_3-based etch recipe which targets SiO_2. Figure 3.35 plots the diameters of nanospheres as a function of etch durations. The etching process also induces dimensional nonuniformity amongst spheres. Initially, the 192-nm-diameter spheres self-assemble into a monolayer with high uniformity (± 2 nm). Once the etch duration exceeds the seventh minute, the sphere shrinks rapidly. The rate of shrinkage increases further after the twenty-fifth minute.

Fig. 3.35 Plot of normalized diameter of nanosphere verses etch duration. Reprinted from Ref. [3] with permission from AIP Publishing LLC

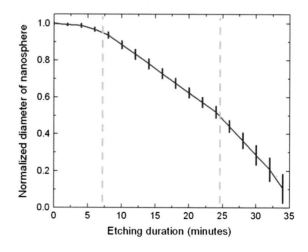

Fig. 3.36 FE-SEM image
demonstrating poor
dimensional uniformity of
spheres due to excessive
etching. Reprinted from Ref.
[3] with permission from AIP
Publishing LLC

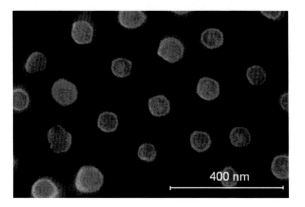

Nevertheless, with increasing etching duration, the variation of diameters between spheres gradually enlarges. When the diameters of spheres are below one-third of their original values, the variation becomes significant and causes poor uniformity (±8), as shown in Fig. 3.36. Therefore, there is a limit on the extent of shrinkage that can be tolerated. When the pattern is transferred to the GaN wafer, nanopillar arrays with poor uniformity of diameters are unfavorable for establishing a well-defined PBG.

To illustrate the effect of nonuniformity qualitatively, finite-difference time-domain (FDTD) simulations were carried out to predict the field distribution when a continuous wave at 440 nm was emitted from the center bottom position from arrays of nanopillars with different packing orders. Figure 3.37a shows the simulation result for an ideal air-spaced nanopillar array with diameter/pitch/height of 140 nm/192 nm/350 nm, whereby lateral propagation of light is obviously suppressed; based on the simulated band diagram a PBG is indeed predicted between ∼436 to 455 nm, correlating well with the FDTD simulated results. However, the presence of defects and nonuniformities disrupts the orderliness of the PhC. As a result, losses and scattering effects are superimposed upon the PBG, the

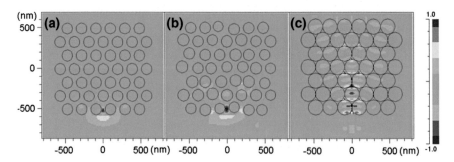

Fig. 3.37 Computed FDTD results for **a** a perfect PhC structure comprising regular air-spaced nanopillars, **b** a PhC structure with defects and **c** a close-packed nanopillar array. Reprinted from Ref. [3] with permission from AIP Publishing LLC

degree of which depends on the extent of defects. A simulation was then performed on a disordered array of nanopillars whereby randomly selected pillars with height of 350 nm have been shifted from their equilibrium positions. At the same time, the diameters of nanopillars in the array range between 130 to 150 nm; such a "defective" array bears close resemblance to an actual fabricated array under worst-case conditions. The results of this simulation are shown in Fig. 3.37b. The introduced defects interrupt the periodic order causing some degree of scattering. The incident source is now partially reflected and partially transmitted by the array due to the presence of leakage modes. This also explains why the measured transmission dips are Gaussian in profile, instead of being abruptly sharp. To complete the picture, an FDTD simulation was also performed on a close-packed 192 nm nanopillar array as illustrated in Fig. 3.37c. The wave is allowed to propagate freely through the array in the absence of a PBG. Comparing the three presented scenarios, one can conclude that the air-spaced nanopillar structure proposed in this work behaves as a leaky PhC, comprising a superposition of defect leakage modes upon PBG confinement modes. In spite of imperfections, its effectiveness on enhancing light extraction from GaN materials is well demonstrated, especially in consideration of the ease and cost of the self-assembled nanoscale patterning process.

3.3.7   Conclusion

In summary, the fabrication of ordered hexagonal array of air-spaced nanopillar on GaN wafers by NSL has been demonstrated. Employing a dual-step dry etch process, the dimensions of the nanopillars in an array can be adjusted without altering their pitch; at the same time, PBG properties of the self-assembled nanostructures can also be modified. The positions of the PBG were identified by optical transmission measurement, correlating well with the prediction of PWE stimulations. A maximum fourfold increase in PL intensity was observed compared to an as-grown sample, depending on the overlap between the PBG with the emission band. Despite the presence of leakage modes due to defects, we have demonstrated the effectiveness of the self-assembled PBG structures in suppressing unwanted guiding modes and promoting light extraction efficiency.

3.4   Tunable Clover-Shaped GaN Photonic Bandgap Structures Patterned by Dual-Step NSL

The fabrication of close-packed clover-shaped photonic crystal structure on GaN by dual-step NSL is demonstrated. By shrinkage of spheres prior to pattern transfer, a non-close-packed clover-shaped PBG structure, as designed by modified 3-D

FDTD simulation, is also realized. The PBG of the close-packed and non-close-packed clover-shaped structures are verified through optical transmission spectroscopy, found to agree well with simulated results. A threefold enhancement in PL intensity is observed from the optimized structure, when the PBG is tuned to overlap with the emission band of the InGaN/GaN multiquantum wells. From time-resolved PL measurements, shortened decay lifetimes are observed.

3.4.1   Introduction

NSL has emerged as a practical approach for patterning large-area-ordered periodic nanostructures. It overcomes resolution issues arising from diffraction limit in optical lithography and even beam size limitations in e-beam lithography. For the standard NSL process, the self-assembled sphere array then serves as an etch mask for pattern transfer to form periodic arrays of recessed air holes or protruding pillars. However, close-packed nanostructures do not provide a suitable periodic variation of refractive index in the lateral direction, so that a PBG corresponding to TE-dominated emission from InGaN/GaN MQWs does not exist. From previous work in this dissertation, air-spaced PBG structures in the visible spectral regions have been demonstrated. The work being presented here adopts a dual-step NSL for generating close-packed (CP) clover-shaped PhC structures. It can also be extended to pattern non-close-packed (NCP) clover-shaped structures through an additional dimension-adjusting process. By adjusting the air spacing, the PBG position can be tuned to spectrally overlap with the emission band of InGaN/GaN MQWs. Transmission measurement are carried out to locate the position of the PBG, which is found to correlate well with 3-D FDTD simulated results. The effects of having a PBG on light extraction and recombination decay lifetime are also discussed.

3.4.2   Process Flow

Figure 3.38 illustrates the process flow for fabricating a clover-shaped PhC structure. The MOCVD grown InGaN/GaN MQW LED on c-plane sapphire substrate emits at center wavelength of $\sim$450 nm and FWHM of $\sim$44 nm. Silica spheres with mean radius of 132 nm are initially diluted in deionized water to produce the optimized volume concentration of cv $\sim$2 %. 5 μL of diluted colloidal suspension mixed with SDS at a volume ratio of 10:1, is dispensed and dispersed uniformly across the sample by spin coating. The SDS surfactant reduces water tension and prevents spheres from aggregating into clusters, thus forming monolayer of spheres. The ordered hexagonal pillar pattern is transferred to GaN by ICP etching using Cl_2/He gas mixtures. The coil and platen powers are maintained at 500 W and

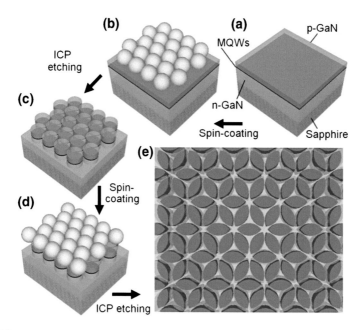

Fig. 3.38 Schematic diagram depicting the process flow; **a** the starting LED wafer; **b** silica spheres coated onto the wafer surface by spin coating; **c** pattern transfer to GaN by ICP etching; **d** second monolayer of sphere array coated on top of pillar array; **e** clover-shaped pattern formed after ICP etching. Reprinted from Ref. [4] with permission from AIP Publishing LLC

135 W while the chamber pressure is held constant at 5 mTorr. The spheres are then removed via sonication in deionized water. The etched sample is subsequently subjected to another NSL process. During spin coating, the spheres spontaneously occupy locations at the triangular voids between adjacent pillars. After another dry etch process, the hexagonal CP clover-shaped PhC is formed. The surface morphologies of the resultant structures are imaged by field emission scanning electron microscope. The FE-SEM images in Fig. 3.39 show the resultant CP and NCP clover-shaped structure, respectively.

Time-integrated PL at room temperature is conducted to characterize the optical properties of PhC structure. The 349-nm excitation laser beam is focused onto the sample while the PL signal is fiber collected to a spectrometer. Time-resolved PL spectroscopy is conducted at room temperature using a picosecond laser (Passat Compiler) as an excitation source, whose wavelength is 266 nm with 8 ps pulse width and 100 Hz repetition rate. The PL signal is band-pass filtered and collected via a 40 × UV objective, subsequently detected by a high-speed photodetector (Thorlabs SVC-FC, < 150 ps rise time), whose electrical signal is read on a 4 GHz digital real-time sampling oscilloscope (Agilent DSO9404A, 85 ps rise time).

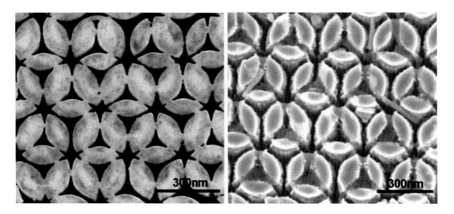

Fig. 3.39 FE-SEM images showing the CP clover-shaped PhC (*left*) and the NCP clover-shaped PhC (*right*). Reprinted from Ref. [4] with permission from AIP Publishing LLC

3.4.3 Photonic Bandgap Calculation

The band structures are computed by modified 3-D- FDTD simulations to predict the PBG position of the clover-shaped PhC structure. The unit cell transformation technique is employed [23], offering compatibility with any geometry, including the clover-shaped structure of this study. Since the TE mode is dominant in emission of InGaN/GaN MQWs, the TE band structures with varying ratios of pillar radius to pitch (*r/a*) are computed for both pillar and clover-shaped arrays, as plotted in Figs. 3.40 and 3.41, respectively. For the pillar arrays, the calculated bandgaps are discretely distributed throughout the plot. Although tuning of PBG can be achieved by modifying *r/a* ratios, the bandwidths are relatively narrow, varying from 8.98 nm at *r/a* = 0.48 (close-packed) to 20.55 nm at *r/a* = 0.41, for effective coupling with QW emission. A detailed study on air-spaced pillar arrays has been reported in previous section. On the other hand, the CP clover structure, with an *r/a* ratio of 0.48, possesses a PBG centered at 517.72 nm and a bandwidth of 12.18 nm. This PBG, being located at the green spectral region does not coincide with the blue light emission from the InGaN/GaN MQWs of the wafers used in this study. However, the computed results also indicate that the PBG shifts gradually toward shorter wavelengths with decreasing *r/a* ratios, together with a broadening of the bandwidth to 53.87 nm at *r/a* of 0.41.

To achieve a reduction of *r/a*, a selective dry etch step [3] can be inserted into the process flow to reduce the sphere radius *r* prior to pattern transfer onto GaN, ensuring that the centroids of the spheres do not shift so that the pitch *a* remains unchanged. Compared to the CP pillar array, spacing between pillars has been established with diameters of pillars reducing to ∼230 nm. The sphere shrinkage process is repeated for the second sphere coating. The resultant NCP clover structure as illustrated in Fig. 3.39 has an *r/a* ratio of ∼0.43, and is predicted to have a complete TE-PBG in the blue spectral region according to Fig. 3.41.

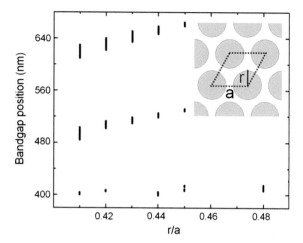

Fig. 3.40 Calculated bandgap as a function of *r/a* for pillar arrays. Reprinted from Ref. [4] with permission from AIP Publishing LLC

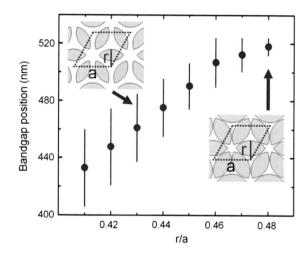

Fig. 3.41 Calculated bandgap as a function of *r/a* for clover-shaped PhCs structures. Reprinted from Ref. [4] with permission from AIP Publishing LLC

3.4.4 Transmission Measurements

To experimentally determine the position of the PBG, an optical transmission measurement is conducted in the near-planar direction. The incident beam from a broadband solid-state plasma light source is focused onto the sample while the transmitted signal is measured by an optical spectrometer. Two pronounced dips with center wavelengths at 517.97 and 447.85 nm are clearly observed in Fig. 3.42 from the transmission spectra of the CP and NCP clover structures, corresponding to their respective bandgap positions. Transmission within the bandgap region does not fall to zero, attributed to out-coupling of light induced by random disorder within the PhCs. Such disorders originate from point and line defects, formed

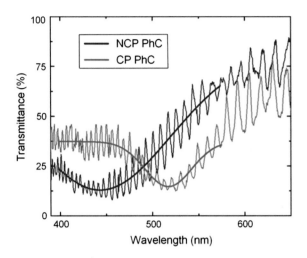

Fig. 3.42 Measured optical reflection spectra from clover-shaped PhCs. Reprinted from Ref. [4] with permission from AIP Publishing LLC

during the coating process, mainly arising from nonuniformities of sphere sizes and geometrical irregularities. At wavelengths beyond the range of frequencies covered by the PBG, relatively high transmission is maintained.

The calculated band structure for the NCP clover structure as shown in Fig. 3.43a also confirms the presence of a PBG at the frequency range a/λ of 0.545 to 0.604, corresponding wavelengths of 437.09 to 484.40 nm. Figure 3.43b shows the narrower PBG located at the green light spectral region of 511.63 to 523.81 nm ($a/\lambda = 0.504$ to 0.516) for the CP array. The measured transmittance data correlate well with simulated results.

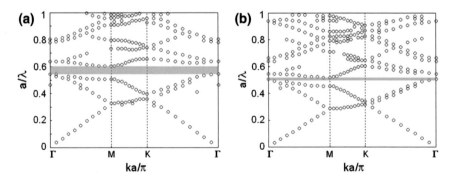

Fig. 3.43 Simulated band diagram of **a** NCP and **b** CP clover-shaped structure, predicting PBG centered at ~517.72 nm and ~460.75 nm, respectively. Reprinted from Ref. [4] with permission from AIP Publishing LLC

3.4.5 Angular-Resolved PL Measurements

The optical enhancement of PhCs incorporation is assessed by conducting an angular-resolved PL measurement on the structures. The integrated PL intensity is collected at 1° intervals over an angular range of 90° at RT. Owing to high refractive index contrast at the GaN/air planar interface, a light beam striking the interface at incident angles beyond the critical angle determined by sin^{-1} (n_{air}/n_{GaN}) $\approx 24.6°$ will remain confined due to total internal reflection (TIR) and sequentially lost due to the reabsorption by the active layer. Hence the rapid drop in PL intensity at angles beyond $\sim 30°$ with respect to the normal, giving a narrow escape cone. According to the angular PL plot in Fig. 3.44, the FWHM of emission divergence increases from 128.72° in the unpatterned sample to 138.26° in the NCP PhC and 137.18° in the CP PhC; the clover-shaped PhCs can indeed expand the escape cone of light. Although the PBG position of the NCP clover structure does not correspond to emission spectrum, the nanotextured surface can still increase the probability of light escaping from the wafer via surface scattering, diminishing the losses caused by TIR. Compared to the unpatterned sample, a threefold increase in PL intensity is observed from the NCP structure. The result indicates that proper design of a PhC is crucial for maximizing light extraction. The approach described here offers the capability of bandgap tuning, allowing optimal overlap between the PBG and emission wavelengths.

To further investigate the emission behavior of the PhCs, the PL spectra at each angle are measured; the individual normalized spectra are combined to generate an angular-resolved emission patterns. For the as-grown sample, sharp Fabry–Pérot interference fringes are observed as shown in Fig. 3.45a, resulting from multiple reflections at the GaN/sapphire and the air/GaN interfaces, both with high refractive index contrasts. The NCP PhC structure maintains uniform emission intensity with faint fringes, as shown in Fig. 3.45b, signifying that the optical confinement effect has been weakened. The improvement is attributed to Brillouin zone folding. Embedding such "weak" PhCs on top of the GaN wafer causes the dispersion curves of Bloch modes to become folded at the Brillouin zone boundary, allowing phase

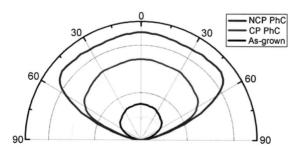

Fig. 3.44 Angular PL plot showing integrated PL intensity versus azimuthal angle. Reprinted from Ref. [4] with permission from AIP Publishing LLC

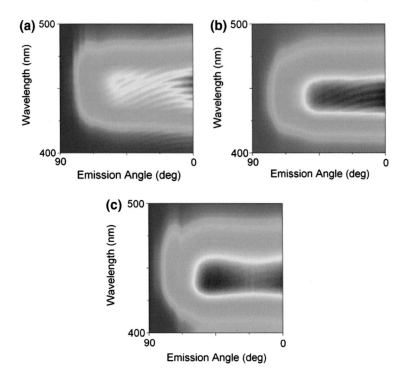

Fig. 3.45 Angular-resolved emission patterns for **a** as-grown, **b** CP and **c** NCP clover-shaped PhCs. Reprinted from Ref. [4] with permission from AIP Publishing LLC

matching to the radiation modes that lie above this cutoff frequency. The PhCs can further Bragg scatter the light emitted from of the active region to avoid Fabry–Pérot oscillations. The CP clover structure, on the other hand, acts as a "strong" PhCs via a different mechanism- the PBG effect which significantly alters the properties of light propagation. The fringe-free spectrum obtained in Fig. 3.45c indicates that lateral propagation of Bloch guided modes is prohibited by the PBG, thereby light generated from MQWs will couple directly to radiation modes. Additionally, the spectrum from the clover structures exhibit a spectral blueshift of ∼6 nm compared to the as-grown, attributed to the partial strain relaxation of the MQW.

3.4.6   Time-Resolved PL Measurements

A time-resolved PL measurement is then carried out to examine the carrier dynamics. The PL decay rate can generally be expressed as the sum of the radiative and nonradiative recombination rates. The experimental data is fitted into two exponential decay profiles, as shown in Fig. 3.46. The fast decay component τ_F, which is strongly dependent upon thermally activated nonradiative recombinations

Fig. 3.46 Time-resolved PL decay profiles measured at room temperature. These decay profiles are fitted by double-exponential decay curves. Reprinted from Ref. [4] with permission from AIP Publishing LLC

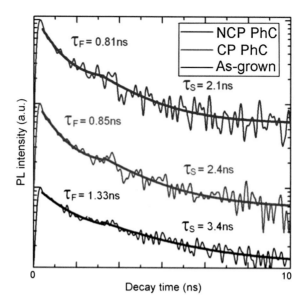

at room temperature, is shortened to 0.39 and 0.51 ns for the NCP and CP PhCs, respectively, from 1.03 ns for the as-grown, attributed to higher surface recombination velocity as a result of increased etched sidewalls. On the other hand, the clover nanostructures indeed enhance carrier localization and reduce the quantum-confined stark effect, thereby accelerating the radiative recombination rates, as evident from the slow decay component τ_S.

3.4.7 Conclusion

The formation of CP and NCP clover-shaped PhCs has been demonstrated by dual-step NSL. The PBG determine from measured optical transmission spectra agree well with the band structure as calculated by modified 3-D FDTD simulations. A threefold enhancement in PL intensity is observed from the NCP clover-shaped PhCs, which has been optimized for the MQW emission band. Shortened PL decay lifetimes observed at room temperatures from PhCs structures suggests nanocavities effects.

3.5 Chapter Summary

In this chapter, the fabrications of four PhCs structures on GaN LEDs by NSL are demonstrated. Silica spheres acting as etch mask are self-assembled into monolayer and transfer pattern to substrate after dry etching. A wide range of dimensions of silica spheres are used to generate two types of well-ordered periodic structures, namely weak and strong PhCs. Both PhCs nanostructures enable strong interaction of the guided modes introduced by total internal reflection, so as to realize enhancement of light extraction of LEDs. Weak PhCs composed of close-packed pillar can extract light from LEDs via dispersion behavior and diffraction property of PhCs while strong PhCs patterned by modified NSL are capable of suppressing undesired optical guiding mode via photonic bandgap and promoting light extraction. The optical performances and mechanisms of weak and strong PhCs are also investigated with the acid of plane wave expansion and FDTD simulations.

References

1. Li KH, Choi HW (2011) InGaN light-emitting diodes with indium-tin-oxide photonic crystal current-spreading layer. J Appl Phys 110(5). Artn 053104. doi:10.1063/1.3631797
2. Li KH, Zang KY, Chua SJ, Choi HW (2013) III-nitride light-emitting diode with embedded photonic crystals. Appl Phys Lett 102(18). Artn 181117. doi:10.1063/1.4804678
3. Li KH, Choi HW (2011) Air-spaced GaN nanopillar photonic band gap structures patterned by nanosphere lithography. J Appl Phys 109(2). Artn 023107. doi:10.1063/1.3531972
4. Li KH, Ma ZT, Choi HW (2012) Tunable clover-shaped GaN photonic bandgap structures patterned by dual-step nanosphere lithography. Appl Phys Lett 100(14). Artn 141101. doi:10. 1063/1.3698392
5. Wiesmann C, Bergenek K, Linder N, Schwarz UT (2009) Photonic crystal LEDs - designing light extraction. Laser Photonics Rev 3(3):262–286. doi:10.1002/lpor.200810053
6. Lee YJ, Kim SH, Huh J, Kim GH, Lee YH, Cho SH, Kim YC, Do YR (2003) A high-extraction-efficiency nanopatterned organic light-emitting diode. Appl Phys Lett 82 (21):3779–3781. doi:10.1063/1.1577823
7. Kisielowski C, Kruger J, Ruvimov S, Suski T, Ager JW, Jones E, LilientalWeber Z, Rubin M, Weber ER, Bremser MD, Davis RF (1996) Strain-related phenomena in GaN thin films. Phys Rev B 54(24):17745–17753. doi:10.1103/PhysRevB.54.17745
8. Rieger W, Metzger T, Angerer H, Dimitrov R, Ambacher O, Stutzmann M (1996) Influence of substrate-induced biaxial compressive stress on the optical properties of thin GaN films. Appl Phys Lett 68(7):970–972. doi:10.1063/1.116115
9. Takeuchi T, Wetzel C, Yamaguchi S, Sakai H, Amano H, Akasaki I, Kaneko Y, Nakagawa S, Yamaoka Y, Yamada N (1998) Determination of piezoelectric fields in strained GaInN quantum wells using the quantum-confined Stark effect. Appl Phys Lett 73(12):1691–1693. doi:10.1063/1.122247
10. Leroux M, Grandjean N, Laugt M, Massies J, Gil B, Lefebvre P, Bigenwald P (1998) Quantum confined Stark effect due to built-in internal polarization fields in (Al,Ga)N/GaN quantum wells. Phys Rev B 58(20):13371–13374
11. Takeuchi T, Sota S, Katsuragawa M, Komori M, Takeuchi H, Amano H, Akasaki I (1997) Quantum-confined stark effect due to piezoelectric fields in GaInN strained quantum wells. Jpn J Appl Phys 2, 36(4A):L382–L385. doi:10.1143/Jjap.36.L382
12. Nishida T, Saito H, Kobayashi N (2001) Efficient and high-power AlGaN-based ultraviolet light-emitting diode grown on bulk GaN. Appl Phys Lett 79(6):711–712. doi:10.1063/1. 1390485
13. Liu L, Edgar JH (2002) Substrates for gallium nitride epitaxy. Mater Sci Eng R 37(3):61–127. Pii S0927-796x(02)00008-6. doi:10.1016/S0927-796x(02)00008-6

14. Wierer JJ, David A, Megens MM (2009) III-nitride photonic-crystal light-emitting diodes with high extraction efficiency. Nat Photonics 3(3):163–169. doi:10.1038/Nphoton.2009.21

15. Fan SH, Villeneuve PR, Joannopoulos JD, Schubert EF (1997) High extraction efficiency of spontaneous emission from slabs of photonic crystals. Phys Rev Lett 78(17):3294–3297. doi:10.1103/PhysRevLett.78.3294

16. Kim HM, Cho YH, Lee H, Kim SI, Ryu SR, Kim DY, Kang TW, Chung KS (2004) High-brightness light emitting diodes using dislocation-free indium gallium nitride/gallium nitride multiquantum-well nanorod arrays. Nano Lett 4(6):1059–1062. doi:10.1021/Nl049615a

17. Shapiro NA, Kim Y, Feick H, Weber ER, Perlin P, Yang JW, Akasaki I, Amano H (2000) Dependence of the luminescence energy in InGaN quantum-well structures on applied biaxial strain. Phys Rev B 62(24):R16318–R16321. doi:10.1103/PhysRevB.62.R16318

18. Ambacher O, Majewski J, Miskys C, Link A, Hermann M, Eickhoff M, Stutzmann M, Bernardini F, Fiorentini V, Tilak V, Schaff B, Eastman LF (2002) Pyroelectric properties of Al(In)GaN/GaN hetero- and quantum well structures. J Phys-Condens Matter 14(13):3399–3434. Pii S0953-8984(02)29173-0. doi:10.1088/0953-8984/14/13/302

19. Yablonovitch E (1987) Inhibited spontaneous emission in solid-state physics and electronics. Phys Rev Lett 58(20):2059–2062. doi:10.1103/PhysRevLett.58.2059

20. Eliseev PG, Smolyakov GA, Osinski M (1999) Ghost modes and resonant effects in AlGaN-InGaN-GaN lasers. IEEE J Sel Top Quantum 5(3):771–779. doi:10.1109/2944.788450

21. Boroditsky M, Krauss TF, Coccioli R, Vrijen R, Bhat R, Yablonovitch E (1999) Light extraction from optically pumped light-emitting diode by thin-slab photonic crystals. Appl Phys Lett 75(8):1036–1038. doi:10.1063/1.124588

22. Jiang HX, Lin JY (2003) III-nitride quantum devices—microphotonics. Crit Rev Solid State 28(2):131–183. doi:10.1080/10408430390802440

23. Ma ZT, Ogusu K (2009) FDTD analysis of 2D triangular-lattice photonic crystals with arbitrary-shape inclusions based on unit cell transformation. Opt Commun 282(7):1322–1325. doi:10.1016/j.optcom.2008.12.055

Chapter 4
Whispering Gallery Mode Lasing from Sphere-Patterned Cavities

Abstract As discussed in Chap. 3, versatile nanopatterns can be created through various smart tweaks of the NSL process, such as applying additional deposition and/or dry etch steps on the HCP nanosphere template. As opposed to enhancing light extraction, the sphere-patterned structures described in this chapter act as optical cavities that support resonant mode in the strong confinement regime, enabling blue/violet lasing at room temperature. Two WG cavities, namely the nano-ring array and metal-clad nanopillar arrays, have been fabricated via modified NSL processes and confine prorogating photons along closed circular paths. The morphologies of the resultant ordered nanostructures and the optical characteristics are discussed.

4.1 Overview

WG mode dielectric resonators have received considerable attention as they offer the potential to pursue high-Q factor and low threshold lasing. The resonator is generally designed as a disk-shaped element to take full advantage of total internal reflections; light is guided by continuous total internal reflection along a curved surface. The pioneering work of WG modes lasing from nitride-based material has been reported by Choi et al. [1]; an array of pivoted GaN microdisks is fabricated on a GaN/Si material. The circular mask pattern with diameters of 20 μm is defined on a GaN-on-Si wafer by photolithography and the microdisk structure is transferred onto the GaN by Cl_2/BCl_3 plasma etching. The sample is then wet-etched in a mixture of HNO_3, H_2O, and NH_4F to create an undercut as this isotropic wet etchant removes the Si surrounding the GaN microdisks. The tiny Si pedestal supports the microdisk and provides a path way for current injection, as revealed in the SEM images in Fig. 4.1a, b. The microdisk structure strongly confines WGM in the disk active layer by total internal reflection at semiconductor/air boundaries. Multiple resonant modes, corresponding to WGM, are observed in PL spectra in Fig. 4.2, with mode spacing of ∼1.1 nm and measured Q factor of 80.

© Springer-Verlag Berlin Heidelberg 2016
K.H. Li, *Nanostructuring for Nitride Light-Emitting Diodes and Optical Cavities*, Springer Theses, DOI 10.1007/978-3-662-48609-2_4

(a)   **(b)**

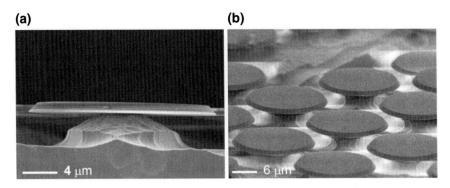

Fig. 4.1 FE-SEM images showing **a** the cross-sectional view and **b** tile-view of pivoted microdisk structure. Reprinted from Ref. [1] with permission from AIP Publishing LLC

Fig. 4.2 PL spectrum of the microdisk at increasing excitation power; the *inset* shows a plot of integrated PL intensity versus excitation energy. Reprinted from Ref. [1] with permission from AIP Publishing LLC

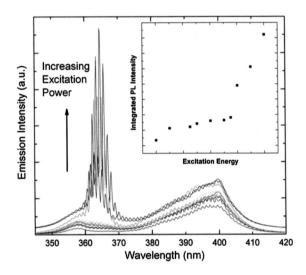

It is worth to note that the frequency and mode spacing is strongly dependent on the dimension of disk-shaped element. Microdisk structures defined by photolithography generally have narrow mode spacing, resulting in dense resonant peaks distributed throughout the spectrum. In order to achieve single WG mode lasing for short visible wavelength and even UV radiation, dimensional downscaling of circular structure is necessary. Nanopillar structures of submicron dimensions would have larger mode spacing, which is suitable for minimizing the number of modes overlapping the gain spectrum, thus achieving single-mode stimulated emission. During the pattern transfer from the nanosphere mask, the resultant circular-shaped pillars are potentially well-suited for establishing WGM. As such, sphere-patterned cavities may be a promising approach due to its ability to produce structures of very small dimensions, as well as high-density integration on a wafer [2, 3].

4.2   UV Single-Mode Lasing from Metal-Clad Pillar

An ordered hexagonal close-packed nanopillar array is fabricated on GaN. A metal coating is then applied to encapsulate the pillars for promoting optical confinement within the cylindrical cavity. Room temperature lasing at 373 nm is observed under pulsed excitation, at a lasing threshold of 0.42 MW/cm^2. With pillar diameters of around 980 nm, the number of modes overlapping with the emission spectrum is reduced, giving rise to single-mode whispering gallery-stimulated emission. FDTD simulations are carried out for the prediction of resonant frequencies and electric field patterns corresponding to the resonant modes.

4.2.1   Fabrication Process

NSL is employed to fabricate a large-area ordered hexagonal nanopillar array and the process flow is described in the previous chapter. Silica spheres with mean diameters of 990 nm are spin-coated onto a nitride-based wafer grown by MOCVD on c-plane sapphire. The periodical nanosphere pattern is transferred to GaN by ICP etching. After sphere mask removal, a metal coating is deposited over the nanopillar array by e-beam evaporation. The coated sample was dry etched for 180 s giving an etch depth and pillar diameter of approximately 750 nm and 980 nm, respectively, evident from Fig. 4.3c. To ensure that the air-gaps between neighboring pillars are fully filled with metal, a higher deposition rate of around 3 Å/s was used. The pillars were almost completely covered by the metal clad, as shown in Fig. 4.3d.

4.2.2   Optical Properties

The optical properties of the nanopillars are evaluated by conducting room-temperature PL measurements, by using the same setup mentioned in the section. Instead of exciting PL from the p-GaN side, the focused beam with spot diameter of 60 μm was incident perpendicularly through the bottom polished sapphire surface since the metal layer strongly attenuates the incident UV beam. The PL signal was collected with a fiber probe at a detection angle of $\sim 80°$ to the normal. Figure 4.4 shows the r.t. PL spectra of four nanopillar cavity samples without and with different metal claddings—namely Au, Ag, and Al—under excitation density of ~ 0.85 MW/cm^2. Two broadband emissions centered at ~ 366.43 and ~ 433.07 nm were observed from the nanopillar sample, corresponding to spontaneous emissions from GaN and InGaN/GaN MQWs. The intensity of MQW emission is much lower than that of GaN emission since the excitation beam was incident through the substrate side. A significant portion of the

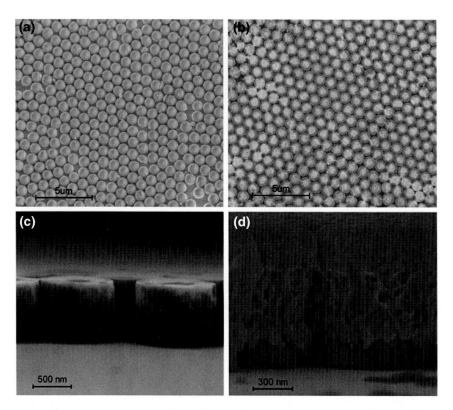

Fig. 4.3 FE-SEM images showing **a** ordered hexagonal monolayer arrays of spheres; **b** pillar array with etch depth and pillar diameter of 750 nm and 980 nm respectively; cross-sectional views of pillars **c** before and **d** after metal coating. Reprinted from Ref. [2] with permission from OSA

Fig. 4.4 Room temperature PL spectra of nanopillars with various metal coatings at high excitation power density of ~ 0.85 MW/cm^2. Reprinted from Ref. [2] with permission from OSA

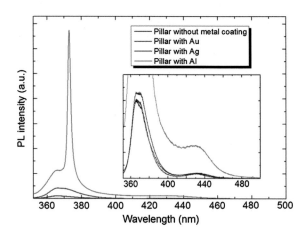

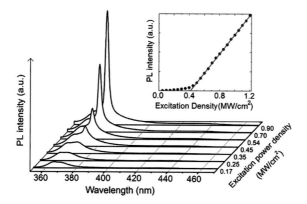

Fig. 4.5 PL spectra of Al-coated nanopillars under varying excitation power density, with lasing mode observed at 373 nm; the *inset* shows integrated PL intensity versus pump power density. Reprinted from Ref. [2] with permission from OSA

incident photons have been absorbed by the GaN layer before propagating to the MQWs region, resulting in domination of GaN UV emission.

To promote resonant modes within the nanopillar, Au, Ag, and Al metal layers are coated onto the surface to encapsulate the cavities. From the PL spectra, a resonant mode was observed only from the Al-coated sample, while the other samples experience mild PL enhancements. The Al coating, with the highest UV reflectivity, is thus demonstrated to be capable of enhancing the confinement effects of a pillar cavity. Figure 4.5 shows excitation-varying PL spectra from the Al-coated sample. Two spectral peaks centered at ∼365 and ∼373 nm can be resolved. However, when the excitation density exceeds threshold (extracted from the plot to be ∼0.42 MW/cm2), the latter peak becomes dominant. The plot in the inset of Fig. 1.15 verifies the nonlinear increase in PL intensity above threshold, characteristics of stimulated emission. Figure 4.6a, b shows captured CCD images of the PL spot below and above threshold respectively. Above threshold, speckle-like patterns are clearly observed, attributed to coherent emission.

While the geometries of nanopillars support WG modes, the presence of a metal coating may also introduce vertical F-P modes. To distinguish between the mechanisms, an Al layer was coated onto an as-grown wafer for comparison. Both the nanopillar and as-grown samples are then subjected to UV laser excitation at 0.85 MW/cm2, the PL spectra of which are plotted in Fig. 4.7, indicating that no optical modes can be observed from the Al-coated as-grown sample; the likelihood of F-P mode lasing is therefore slim. With a single mirror coating, photons can still escape from the opposite GaN/sapphire or sapphire/air interfaces, resulting in inefficient optical confinement. On the other hand, pillar cavities with metal coating on both the sidewall and top surface provide strong lateral confinement within the circular cavity. This is particularly so for Al, which provides a high reflectivity in the UV spectral region, according to the reflectivity curves for the corresponding metals shown in the inset of Fig. 4.7. For reference, with a lower UV reflectivity,

(a)   **(b)**

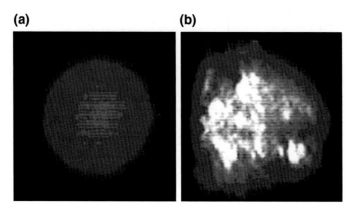

Fig. 4.6 CCD-captured images of PL emission from Al-clad pillar below (*left*) and above threshold. Reprinted from Ref. [2] with permission from OSA

Fig. 4.7 Effects of Al-coating on as-grown and nanopillar samples; the *inset* shows the reflectance curves for Au, Ag, and Al. Reprinted from Ref. [2] with permission from OSA

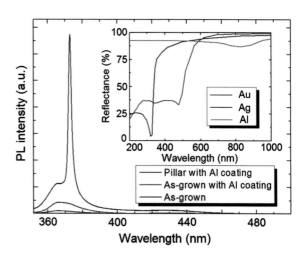

the Ag-clad nanopillars would also lase at a higher threshold of 0.97 MW/cm2. The WG resonator frequency and mode spacing of circular cavities are evaluated by the relations

$$2\pi Rn = m\lambda \tag{4.1}$$

and

$$\Delta\lambda_{WG} = \frac{\lambda^2}{2\pi Rn} \tag{4.2}$$

where R is the pillar radius (490 nm) and n is the effective refractive index (2.3). The lasing peak extracted from the plot is 373 nm, corresponding to m = 19, while the adjacent modes of m = 20 and m = 18 are calculated to be 354 and 393.4 nm, respectively, with mode spacing of around 19.6 nm. Since only the resonant mode at 373 nm overlaps strong GaN emission, single WG mode resonance is observed.

4.2.3   FDTD Simulation

Finite-difference time-domain (FDTD) simulations are performed to verify the resonant characteristics of the nanopillars with different metal claddings. The computational area and the mesh size (Δx) are fixed at $1.5 \times 1.5~\mu m^2$ and 5 nm, respectively, with the time step set to 0.00967 fs satisfy the Courant stability condition:

$$\Delta t_{max} = \frac{\Delta x}{c\sqrt{2}} = 0.0118 fs \qquad (4.3)$$

The simulated Transverse Electric (TE) mode spectra for the Al, Ag, and Au cladded nanopillars in the wavelength range of 400–500 nm are plotted in Fig. 4.8. In the spectrum of the Al-clad pillars, a first-order mode at ~ 365 nm and a higher-order mode at ~ 373 nm are observed, overlapping with the peak of the GaN emission spectrum. Also note that its modal intensities, thus optical confinement, remain high across the frequency range of interest, while that for Ag-clad pillars decline with reducing wavelengths, whereas no resonant modes are observed for the Au-clad pillars. This is consistent with the reflectivities of the respective metal coatings as plotted in the inset of Fig. 4.7. The FDTD-simulated steady-state electric field patterns corresponding to the resonant modes at 365 and 373 nm are shown in Fig. 4.9, neglecting coupling effects. It can be seen that the field

Fig. 4.8 FDTD simulated spectra showing resonant characteristics of metal-clad nanopillar cavities. Reprinted from Ref. [2] with permission from OSA

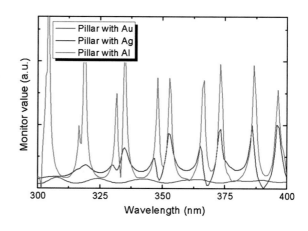

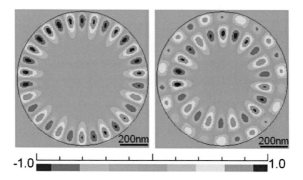

Fig. 4.9 FDTD-computed steady-state electric field patterns of the resonant modes at (*left*) 365 nm and (*right*) 373 nm. Reprinted from Ref. [2] with permission from OSA

corresponding to the first-order mode at 365 nm is concentrated around the periphery of the pillar, and therefore would be prone to scattering losses along the roughened plasma-etched sidewalls. On the other hand, the field pattern for the higher-order mode at 373 nm is established within the pillar and will not suffer from optical losses, explaining why this mode becomes the dominant lasing mode above threshold.

4.2.4 Angular-Resolved PL Spectroscopy

With WG emission modes, a highly directional emission pattern can be expected in the planar directions [4]. To verify this, angular-resolved PL in the near-planar direction is measured from a pillar array by rotating a fiber around the central axis of the sample, with the collection angle maintained at ∼80° to the normal, as illustrated in Fig. 4.10 (top). A high degree of emission directionality is clearly present, with multiple intensity peaks at various discrete angles, deduced from the

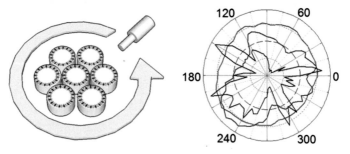

Fig. 4.10 (*Left*) Schematic diagram showing the setup for measuring in-plane angular emission pattern; (*Right*) Angular emission patterns of Al-clad pillars from the in-plane (*black*) and vertical (*blue*) directions. The *curve* in *red* corresponds to in-plane measurement from an as-grown sample. Reprinted from Ref. [2] with permission from OSA

curve in black from the polar plots of Fig. 4.10 (bottom). As the emission modes are scattered out of the cavities through roughened (etched) sidewalls, the recorded pattern is asymmetrical; geometrical imperfections of individual pillars and nonuniformities between pillars adds on to the asymmetry. Such periodicity is clearly absent from either the angular-resolved PL pattern as measured in the near-normal direction (blue curve) or from the as-grown sample (red curve). A final note on the cold-cavity quality factor Q, which has been evaluated as 150 at an excitation power density of 0.4 MW/cm^2. The relatively low magnitude may be attributed to sidewall scattering losses, intra-cavity coupling, and most importantly, the absence of a gain region. However, the low gain is merely a limitation of the PL measurement geometry. In the electroluminescent device under development, emission from the MQWs will be dominant, providing sufficient gain to overcome the losses. Nevertheless, the present optical-pumped demonstration verifies the feasibility of this approach for forming optical micro-cavities.

4.2.5 Conclusion

The fabrication of metal-clad nanopillar structures by NSL is demonstrated. Pillars with Al coating provide strong lateral confinement and room temperature lasing at 373 nm has been observed, at a lasing threshold of 0.42 MW/cm^2. The lasing mechanism is identified as WG mode. Single-mode lasing is possible due to the submicron dimensions of individual pillars.

4.3 High-Q WG Mode from Nanoring Array

A hexagonal-close-packed ordered array of nano-rings was fabricated on GaN with a modified nanosphere lithography process. The spheres initially served as etch masks for the formation of close-packed nanopillars. The spheres were then shrunk and with a layer of oxide deposited, the roles of the spheres became masks for lift-off. The final etch produced nano-rings with wall widths of 140 nm. Photo-pumped lasing with splitting modes was observed at room temperature, with a low lase threshold of ∼10 mJ/cm^2 and high quality factor of ∼5000, via whispering gallery modes. The resonant frequencies were verified through finite-difference time-domain simulations.

4.3.1 Process Flow

As demonstrated in the previous chapters, modification of the fabrication flow can possibly lead to the formation of novel nanoscale geometries. Based on such

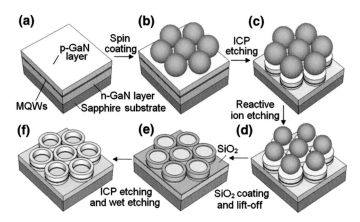

Fig. 4.11 Schematic diagram illustrating the process flow: **a** the starting LED wafer; **b** silica spheres coated on p-GaN layer; **c** pattern transfer to GaN by ICP etching; **d** shrinkage of spheres using RIE etching; **e** SiO_2 coating followed by lift-off; **f** removal of SiO_2,giving final structure. Reprinted from Ref. [3] with permission from AIP Publishing LLC

strategies, we have developed nano-ring arrays supporting WG mode, with high-Q factor and low lasing threshold. The fabrication of the nano-ring arrays evolves from standard processes for forming closed-packed nanosphere monolayers, with additional steps for tweaking the dimensions. Figure 4.11 illustrates the entire process flow for the fabrication of closed-packed nano-ring arrays by NSL. The LED wafer used contains InGaN/GaN MQWs grown by MOCVD on c-plane sapphire substrate, with emission center wavelength of ~ 465 nm.

Silica spheres with mean diameters of 1000 nm were self-assembled on the wafer. The coated monolayer of spheres, which acted as an etch mask, was subsequently transferred to the wafer forming nanopillars by ICP etching using Cl_2 as etchant gas at flow rate of 30 sccm. The spheres were subsequently shrunk in diameters though RIE process using CHF_3-based plasmas under low RF power conditions to minimize lateral motion; this non-close-packed sphere array now serves as a lift-off mask. By electron-beam evaporation, an SiO_2 layer (150 nm) was deposited over the entire surface. Following lift-off by sonication in deionized water, a ring of SiO_2 was left behind at the periphery of the nanopillars, the width of which depends on the extent of the shrinking in the previous step. These oxide rings acted as a mask during the second ICP etch, resulting in the inner-hole region being etched, and the formation of nano-rings. The residue was removed by immersing the wafer into 10 % hydrofluoric (HF) acid. Surface morphologies of the nano-ring arrays were imaged using a Seiko Nanopics atomic force microscope (AFM). The AFM images in Fig. 4.12 revealed assembled nano-ring array with internal diameters of 720 nm, wall widths of 140 nm and heights of 500 nm. The arrays are regularly close-packed in a hexagonal arrangement.

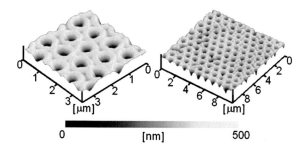

Fig. 4.12 AFM images of the hexagonal-close-packed nano-ring arrays. Reprinted from Ref. [3] with permission from AIP Publishing LLC

4.3.2   Lasing Characteristics

The optical characteristics were evaluated by time-integrated PL at room temperature. A DPSS UV laser at 349 nm emission and pulse duration of 4 ns at 1 kHz repetition rate was used as an excitation source. The focused beam was incident nearly perpendicular to the sample surface with a spot diameter of around 150 μm. The PL signal was collected at 30° to the normal with an optical fiber bundle, and the signal coupled to a spectrometer comprising a spectrograph and a CCD, which offered optical resolutions of better than 0.04 nm. PL spectra of the nano-rings at increasing excitation densities are plotted in Fig. 4.13. Under low excitation, the spectrum exhibit weak spontaneous emission from MQWs at $\lambda = 460$ nm with FWHM of 34 nm. As the excitation energy density exceeds the threshold of 16.8 mJ/cm², a sharp spectral peak emerged at 444 nm; beyond 305.4 mJ/cm², a second spectral peak at 472 nm was observed. Zooming in on the spectrum, three satellite spectral peaks at 443.88, 444.08, and 444.26 nm with FWHM as low as 0.09 nm could be resolved as evident in Fig. 4.14; their quality factors are evaluated as 4455, 5025, and 4660 respectively.

Fig. 4.13 Room temperature PL spectra of nano-ring array under varying excitation energy density. Reprinted from Ref. [3]

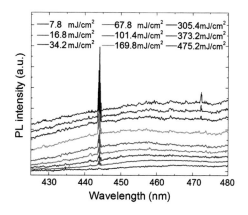

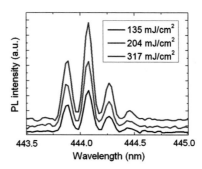

Fig. 4.14 An enlarged view of the PL spectrum near the lasing frequency. Reprinted from Ref. [3] with permission from AIP

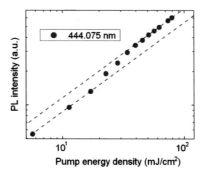

Fig. 4.15 PL intensity of the lasing peak at 444.075 nm as a function of pump energy density in logarithmic scale. Reprinted from Ref. [3] with permission from AIP Publishing LLC

Figure 4.15 plots intensities of the 444.075 nm lasing peak as a function of excitation energy density in logarithmic scale. Nonlinear increases in PL intensities were observed beyond the extracted threshold value of $\sim$10 mJ/cm^2. To characterize the optical cavity, the spontaneous emission coupling factor β, which represent a figure of merit for the cavity, is evaluated. β defined as the fraction of total spontaneous emission from the source that is coupled into a cavity mode, can be estimated from the ratio of the output intensity below and above the lasing threshold [5]. In practice, it is always less than unity due to the concurrent emission of nonresonant modes. β of spectral peak at 444.075 nm was estimated to be 0.75. The Purcell factor, which quantifies the ratio of spontaneous emission rate in a resonant cavity mode to the spontaneous emission rate in the absence of the cavity, can be calculated by equation:

$$F_p = \frac{\beta}{1 - \beta} \qquad (4.4)$$

from which F_p was estimated to be 3. Being greater than unity, it is implied that the spontaneous emission rate has indeed been enhanced by the cavity and that excitons can radiate much faster in the cavity than in free space. To identify the position of WGM lasing frequency, the resonant frequency and mode spacing for a ring cavity were calculated from Eqs. (4.1) and (4.2), where n is the refractive index of GaN, R the ring radius. The calculated mode at $m = 17$ corresponds to the observed lasing peak at 444 nm while the mode spacing is evaluated to be in the range of 26–29 nm, correlating well with the PL spectra.

The mechanism for stimulated emission from the nano-ring array is investigated. F-P mode lasing requires the existence of a longitudinal cavity for light confinement [6, 7]. Without mirror coatings such a cavity would not exist and F-P type lasing can thus be ruled out. On the other hand the circular geometry of the nanostructures hints at the possibility of WGM resonance in the planar direction. Indeed, nano-rings cavities are known to support optical confinement modes which are guided by total internal reflections circling around the peripheries of the rings. By employing effective index method, simulation of a cut-through of the material illustrates that the simulated optical-mode propagation at the wavelength of 444 nm is well confined with the ring with width of 140 nm and height of 500 nm, as shown in Fig. 4.16. This set of data verifies that the mode is well guided and localized in ring wall region.

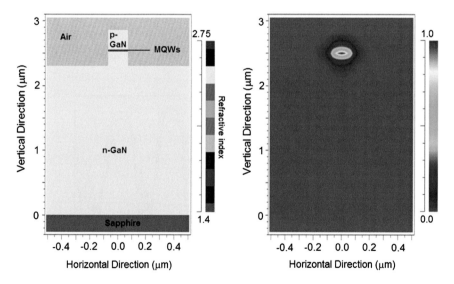

Fig. 4.16 Simulation of a cut-through of the structure by the effective index method

4.3.3   FDTD Simulation

To verify resonant properties of the nano-ring cavity, FDTD simulations are performed on a ring structure with internal diameters of 720 nm and wall widths of 140 nm. To yield better accuracy, the FDTD mesh size is set to less than 1/20 of wavelength. The computational grid contains 150 × 150 nodes over an area of 3 × 3 μm^2, and the time step set to 0.045 fs, sufficiently small to satisfy the Courant stability condition. The simulated transmittance spectra for three different cell structures are shown in Fig. 4.17, which correlate well with the calculated resonant frequencies and mode spacing, and also the lasing positions from the PL spectra. Compared to a single-ring cell, the optical confinement efficiencies of the double-ring and triple-ring cells are increased, implying that multi-coupled ring arrays have stronger optical confinement effects. Apart from such enhancements, splitting of the resonant modes is also introduced for the multi-coupled ring structure [8]. As individual ring resonators possess the same optical path length, no phase-shift changes exist within a single ring. On the other hand, due to strong optical coupling between rings, there exists round-trip phase shifts when light passes through the couplers, thus resulting in the optical splitting effect [9]. It explains why multiple-peaks appear in the vicinity of 444 nm. Additionally, sidewall-roughness induced back-reflection into the counter-propagating modes may yet be another mode-splitting factor [10]. The performance of traveling-wave resonators is affected when the forward- and backward-propagation modes are phase-matched. Such contra-directional coupling leads to annihilation of directionality of a traveling-wave resonator and causes the resonant peak to split into multiple modes. The simulated results shown in Fig. 4.18 illustrate the formation of

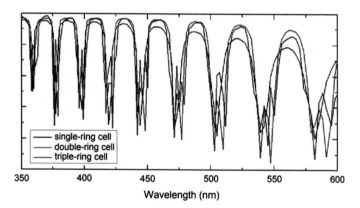

Fig. 4.17 FDTD simulated transmittance plot showing resonant modes for three different cell structures. Reprinted from Ref. [3] with permission from AIP Publishing LLC

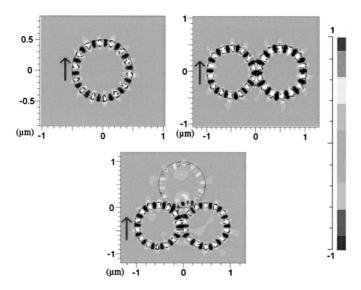

Fig. 4.18 Simulated results showing strong optical coupling occurs between adjacent rings in an array. Reprinted from Ref. [3] with permission from AIP Publishing LLC

standing waves in single-ring structure when a continuous wave at 444 nm was injected at the leftmost position indicated by the red arrow. At the resonant frequencies, photons are effectively confined within the ring structures establishing a WG mode for the single-ring cavity. For the double-ring and triple-ring cells, high-intensity standing waves are set up across the multi-coupled rings structure, indicating that strong light coupling between adjacent rings occurs, and a resonance mode was established due to constructive interference. The ring can thus serve as a resonator and interact further with adjacent rings. This supports the postulation of WG mode lasing observed from PL spectra.

4.3.4   Conclusion

The fabrication of nano-ring arrays by a modified NSL process is demonstrated. The nano-ring network functioned as resonators which support whispering gallery modes. Moreover, room temperature lasing has been observed, with lasing spectral peaks at 443.88, 444.075, and 444.26 nm. Their corresponding Q factors were evaluated to be 4455, 5025, and 4660 respectively. Optical resonance of the nano-ring structure was further verified with FDTD simulations, which was found to correlate well with the frequencies of lasing.

4.4 Chapter Summary

In this chapter, we have described and highlighted two WG mode cavities achieved with nanosphere patterning. Disk-shaped microcavities, namely metal-clad pillar arrays and nano-ring arrays, can effectively confine short-wavelength of light within the circular resonators and demonstrate the blue/UV lasing at room temperature. The size of cavities is sufficiently small to achieve high Q and single-mode lasing for short-wavelength emission. FDTD simulations are also employed to investigate the optical properties.

References

1. Choi HW, Hui KN, Lai PT, Chen P, Zhang XH, Tripathy S, Teng JH, Chua SJ (2006) Lasing in GaN microdisks pivoted on Si. Appl Phys Lett 89(21). Artn 211101. doi:10.1063/1. 2392673
2. Li KH, Ma ZT, Choi HW (2012) Single-mode whispering gallery lasing from metal-clad GaN nanopillars. Opt Lett 37(3):374–376
3. Li KH, Ma ZT, Choi HW (2011) High-Q whispering-gallery mode lasing from nanosphere-patterned GaN nanoring arrays. Appl Phys Lett 98(7). Artn 071106. doi:10. 1063/1.3556281
4. Apalkov VM, Raikh ME (2004) Directional emission from a microdisk resonator with a linear defect. Phys Rev B 70(19). Artn 195317. doi:10.1103/Physrevb.70.195317
5. Takashima H, Fujiwara H, Takeuchi S, Sasaki K, Takahashi M (2008) Control of spontaneous emission coupling factor beta in fiber-coupled microsphere resonators. Appl Phys Lett 92(7). Artn 071115. doi:10.1063/1.2884329
6. Chen R, Sun HD, Wang T, Hui KN, Choi HW (2010) Optically pumped ultraviolet lasing from nitride nanopillars at room temperature. Appl Phys Lett 96(24). Artn 241101. doi:10. 1063/1.3449576
7. Jewell JL, Harbison JP, Scherer A, Lee YH, Florez LT (1991) Vertical-cavity surface-Emitting lasers—design, growth, fabrication, characterization. IEEE J Quantum Electron 27(6):1332–1346. doi:10.1109/3.89950
8. Guider R, Daldosso N, Pitanti A, Jordana E, Fedeli JM, Pavesi L (2009) NanoSi low loss horizontal slot waveguides coupled to high Q ring resonators. Opt Express 17(23):20762–20770. doi:10.1364/Oe.17.020762
9. Smith DD, Chang H, Fuller KA, Rosenberger AT, Boyd RW (2004) Coupled-resonator-induced transparency. Phys Rev A 69(6). Artn 063804. doi:10.1103/ Physreva.69.063804
10. Little BE, Laine JP, Chu ST (1997) Surface-roughness-induced contradirectional coupling in ring and disk resonators. Opt Lett 22(1):4–6. doi:10.1364/Ol.22.000004

Chapter 5
1-µm Micro-Lens Array on Flip-Chip LEDs

Abstract The fabrication of hexagonally close-packed micro-lens array on sapphire face of flip-chip bonded LED by nanosphere lithography is demonstrated. Self-assembled silica spheres serve as an etch mask to transfer hemispherical geometry onto the sapphire. The optical and electrical properties are evaluated. Without degrading the I–V properties, the lensed LED shows an enhancement of 27.8 % on light output power, compared with unpatterned LED. The emission characteristic is also investigated by performing finite-difference time-domain simulation, which is found to be consistent with the experimental results.

5.1 Introduction

III-nitride light-emitting diodes (LEDs) are typically packaged in a way that the sapphire substrates are bonded to a lead frame and light is extracted through the top p-GaN surface. However, the limited conductivity of p-GaN [1, 2] necessitates the coating of a semi-transparent and amorphous indium-tin-oxide current spreading layer, while the insulating sapphire layer makes lateral current injection inevitable. The ITO layer, together with dual electrodes and bond wires, compromises upon device optical performance and emission homogeneity. The flip-chip configuration [3, 4] offers an elegant solution, enabling the extraction of light through the entire transparent and crystalline sapphire substrate. The relatively low contrast of refractive index at the sapphire-air interface enlarges the critical angle for total internal reflections, thereby enhancing light extraction efficiency; direct contact between the light-generating (and this heat-generating) GaN layers with the lead frame promotes effective heat dissipation, improving internal quantum efficiency and device reliability. Flip-chip packaging also offers potential use of the sapphire substrate, onto which optical elements can be monolithically integrated to further boost light extraction or manipulate emission characteristics.

The integration of micro-lens array on LEDs [5, 6], being one of the promising approaches, has been demonstrated to realize the enhanced optical performance of

© Springer-Verlag Berlin Heidelberg 2016

K.H. Li, *Nanostructuring for Nitride Light-Emitting Diodes and Optical Cavities*, Springer Theses, DOI 10.1007/978-3-662-48609-2_5

LEDs. Traditionally, tedious processes involving photo-lithography and thermal resist reflow step are employed to generate micro-lens structures of the order of microns. In this work, we propose the fabrication of hexagonally close-packed micro-lens array on a flip-chip bonded LED via NSL. This self-assembly method of fabricating micro-lens array indeed offers several distinct advantages over conventional photolithographic technique: (i) NSL is the one-step, low cost, and high yield approach to directly define large-area nanopatterns; (ii) Compared with photo-/thermal resist, silica spheres provide relatively high etch resistance (comparable to sapphire), resulting in larger height-to-diameter ratio of lens geometry and more pronounced surface curvature; (iii) inevitably, additional reflow step for reshaping the coated resist into pedestals introduces large spacing while the self-assembly method produces ultimate high-density of HCP hemispherical elements; (iv) NSL, with wide range of scalable sphere dimensions, overcomes the resolution limitation arising from optical diffraction limit. The monolayer of silica spheres acts as an etch mask for transferring pattern onto the sapphire substrate. The lens structures are patterned on the sapphire face of a flip-chip bonded LED, with lens diameters as small as 1 μm [7]. The electrical and optical characteristics of the packaged devices are evaluated.

5.2   Experimental Details

Schematic diagram illustrating the process flow of LEDs is shown in Fig. 5.1. The blue light LED with central wavelength of 440 nm is grown by MOCVD on c-plane sapphire substrate. A semi-transparent ITO film (~ 200 nm) serving as current spreading layer is deposited on top of p-GaN. The sapphire substrate is lapped and polished down to ~ 80 μm. The NSL coating conditions have been previously reported in Chap. 2. Silica spheres with a diameter of 1 μm are diluted in DI water until it reaches an optimized concentration of cv ~ 2 %. 5 μL of droplet is pipetted onto the sapphire surface and uniformly spread into HCP monolayer array by spin-coating process. The self-assembling spheres serve as the etch mask to transfer pattern into sapphire by ICP etching with a gas mixture of SF_6/He. The coil and paten powers are maintained at 400 W and 100 W under a constant pressure of 5 mTorr. During dry etching, the spheres shrink continuously in both lateral and vertical directions, generating hemispherical geometries. Then, the mesa region of 500×200 μm^2 is defined by a standard photolithography process and Cl_2-based plasma is used to expose n-GaN surface. The contact regions are defined by another photolithography and metal pads are coated by e-beam evaporation. A metal lift-off process is followed by rapid thermal annealing to form the ohmic contacts. The chips are flip-chip bonded onto a ceramic sub-mount, as illustrated in the schematic diagram shown in Fig. 5.1e. An unpatterned LED with identical dimension is

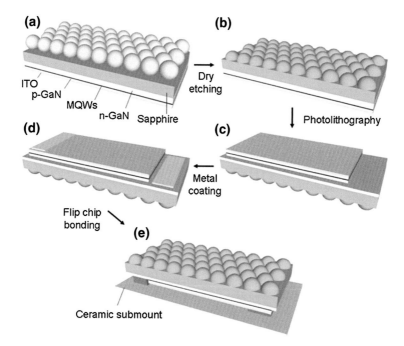

Fig. 5.1 Schematic diagram depicting the process flow; **a** silica spheres coated onto the LED wafer surface by spin-coating; **b** hemispherical patterns formed after ICP etching; **c** mesa definition via photolithography; **d** metal deposited onto pre-defined n-pad and p-pad region; **e** flip-chip bonded onto a ceramic submount

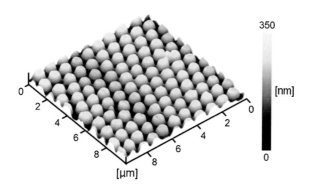

Fig. 5.2 AFM image showing the ordered hexagonal close-packed lens array. Reprinted from Ref. [7] with permission from AIP Publishing LLC

fabricated alongside for comparison. The surface morphology is measured using AFM. The AFM image shown in Fig. 5.2 clearly illustrates the HCP lens array patterned on sapphire substrate, with lens diameter and height of 1 μm and 350 nm respectively.

5.3 Device Characterizations

The optical and electrical properties of lensed LED are evaluated. L–I–V plot shown in Fig. 5.3 illustrates that the curve of lensed LED is similar to that of as-grown LED and their slopes in linear region are almost the same, indicating that the processing does not deteriorate the electrical behavior. At an injection current of 20 mA, the forward voltages of the LEDs with and without lenses are 3.41 and 3.34 V respectively. The electroluminescence measurements are performed on the flip-chip bonded and un-encapsulated devices. The operating devices are measured by a 12-inch integrating sphere coupled to a calibrated spectrometer via an optical fiber. Figure 5.3 also illustrates that the light output power of the lensed LED is enhanced by 27.8 % at an injection current of 90 mA over an unlensed LED. Obviously, the optical enhancement is mainly contributed to the reduced total internal reflection at the sapphire/air interface. In other words, the lensed surface can partially release the trapped photons and lower the reabsorption loss in the active layer.

To examine this phenomenon, a simulation based on RCWA is conducted. The calculated electromagnetic fields as a sum over coupled waves are obtained by solving Maxwell's equations in Fourier domain. The resultant reflectivity is the sum of the reflected diffraction efficiency of all modes. Note that the infinite periodic boundary condition is implemented in the calculation and the unit cells shown in Fig. 5.4 are assumed to be extended infinitely along the x–y plane. Since the flip-chip LEDs contain MQWs emitting blue light with a center wavelength of 440 nm and FWHM of ~ 23 nm, the reflectivities under a sufficient frequency range of 400–500 nm are computed to ensure the emission spectra are fully covered and considered. Obviously, a sharp contrast emerging at $\sim 34.6°$ is observed in Fig. 5.5a. When light rays strike the flat-top surface at an angle larger than 34.6°, its

Fig. 5.3 L–I–V characteristics of the flip-chip LEDs with and without lens array. Reprinted from Ref. [7] with permission from AIP Publishing LLC

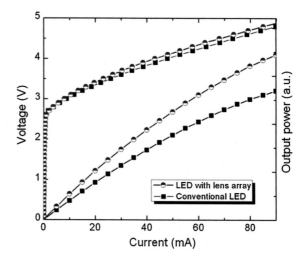

Fig. 5.4 Defined unit cells of periodic lens array and flat-top structure for RCWA simulation. Reprinted from Ref. [7] with permission from AIP Publishing LLC

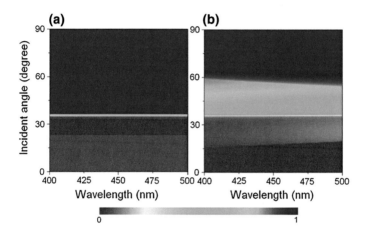

Fig. 5.5 Computed reflectivities of **a** flat-top profile and **b** lens array; linear scale bar corresponds to the reflectivity of incident light. Reprinted from Ref. [7] with permission from AIP Publishing LLC

reflectivity instantly rises to a hundred, which corresponds to the critical angles at the interface between sapphire and air. Also, it is found to be consistent with the Snell's law and the angle can be determined by $\sin^{-1}(n_{air}/n_{sap}) = 34.2°$, where n_{air} and n_{sap} are the refractive indexes of air and sapphire. On the other hand, the lensed sapphire significantly weakens the overall reflectivity over a range of incident angles up to $\sim 58°$, thus enlarging the light-escape cone of LED and allowing more light to escape into free space.

5.4 Far-Field Angular-Resolved Measurement

Apart from the enhancement of light output power, the lens array with convex curvature is predicted to have the ability of altering emission divergence. To investigate this behavior, a three-dimensional FDTD simulation is carried out. FDTD offers the direct solution to Maxwell's equations and presents in a realistic way that the propagating waves vary with respect to the time domain. To have a visible view of the simulated result and to reduce the computational load, instead of constructing the entire devices with hundreds of microns of thickness, a continuous wave (CW) at 440 nm is launched through the scaled-down sapphire to observe the field patterns after propagating through the sapphire/air boundary. Point detectors are placed over emission sector spanning 0°–180° to record the intensities. The spatial resolution in the computational domain is 20 nm (1/22 of wavelength) and the time step is set to be 0.033 fs so as to satisfy the Courant stability condition. To ensure the detected intensities reach steady readings in sufficient time period, the execution time is set at 50 ns. Figure 5.6 shows the simulated results of waves propagating through the lensed and unlensed sapphires. For the flat-top sapphire, emitted waves are uniformly distributed with the wide emission angle, as shown in Fig. 5.6a. The micro-lens integrated sapphire optics condenses the wave upwards, thereby increasing the detected intensities at angles close to the normal direction of LEDs.

Simulated angular plot shown in Fig. 5.7 further quantitatively explains the emission divergence of LEDs. The half-angle of emission, defined as the angle at which power intensity is equal to $1/\sqrt{2}$ of its maximum value and calculated from the lensed sapphire, is reduced by 22.72°, compared with unpatterned structure. To experimentally determine if the lensed LED has any effect on emission divergence,

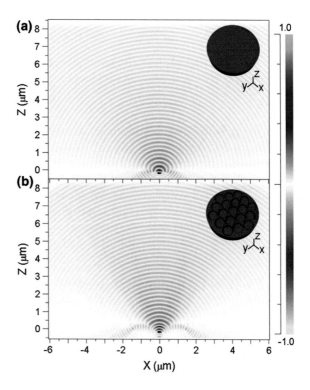

Fig. 5.6 FDTD simulated results of wave propagating through **a** unlensed and **b** lensed sapphire

Fig. 5.7 Calculated angular plot of lensed and bare sapphires

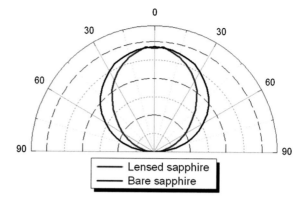

an optical fiber probe coupled to a spectrometer is rotated along radial axis of the LEDs operated at 10 mA in steps of 1°, as illustrated in the schematic diagram shown in Fig. 5.8a. The separation between the fiber probe and the LED is fixed at 5 cm. Normalized polar plot shown in Fig. 5.8b illustrates that the lensed LED produces a reduction of half-angle by 20.2° over an unlensed LED, correlating well with the simulated result. Slight deviation may be due to the existence of point and

Fig. 5.8 **a** Schematic diagram showing the experimental setup for angular emission measurement; **b** measured angular plot of packaged LEDs under an injection current of 10 mA

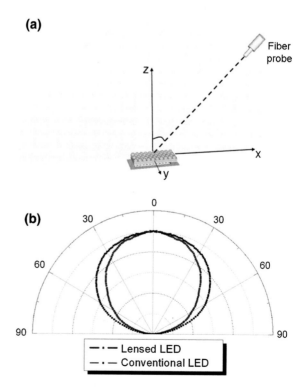

line detects intrinsically formed during self-assembly of spheres. The convergence of emission originates from the focusing effect of the lens array but it is worth noting that the degree of angular divergence is highly dependent on the dimensions of lenses.

5.5 Near-Field Emission Patterns Captured by Confocal Microscopy

As the lens array can enhance the probability of light extraction, the emission profiles of the lensed LEDs are then acquired using confocal microscopy (Carl Zeiss LSM700) to study whether the lens array can alter the directionality of propagating light. Figure 5.9a shows the schematic diagram of the confocal microscopy system comprising of an objective lens, a size-tunable pinhole, and a detector (photomultiplier tubes). Theoretically, the light rays coming from the focal plane can be detected only while the out-of-focus light is effectively suppressed by the size-defined pinhole. After planar confocal images in x–y plane are continuously captured along the z-axis, the cross-sectional view of the combined confocal images can be reconstructed by stacking those captured slices. An external positioning sensor is implemented at the z-axis to monitor the motion of motorized stage and to offer precise axial step of 0.01 μm. The emission pattern of the individual 1-μm lens elements is acquired using a confocal microscope equipped with a ×150 (N.A. = 0.95) objective lens and a blue collimated laser beam is incident on the unpackaged devices in the perpendicular direction on to the planar surface. The

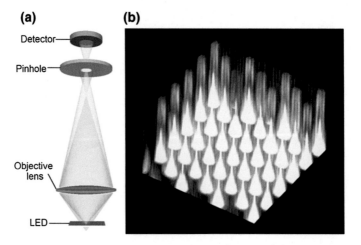

Fig. 5.9 a Schematic diagram illustrating the confocal microscopy system; **b** three-dimensional view of the combined confocal images. Reprinted from Ref. [7] with permission from AIP Publishing LLC

detected signals are slightly overexposed so as to express a more apparent view of focusing properties. Figure 5.10a, b shows the planar view of confocal images recorded at the device surface and the focal plane of lens array. To highlight the focusing phenomenon of the lenses, the cross-sectional view of the stacked confocal images is reconstructed, as illustrated in Fig. 5.10c. The collimated beams are condensed into a spot at a distance of 0.670 μm from the sapphire surface. The focal length (f) of a lens can also be verified by $f = (h + r)/(2\ hn_{sap} - 2h) = 0.682$ μm, where h and r are the radius and height of lens. The theoretical calculation is in good agreement with the value measured by the confocal microscopy. Figure 5.9b shows the comprehensive three-dimensional view of the combined focal images, revealing that the focusing properties are consistently observed from the ordered hexagonal lens array. Although an objective lens with lower magnification can be commonly used to obtain a larger field of view, the captured cone of the objective governed by the decreased value of N.A. is strictly limited and becomes an important factor. To investigate the effect of lens array on the emission profiles of devices, an objective lens with ×20 (N.A. = 0.5) is utilized to provide sufficient view to capture the half size of the operating device and detect the diverse emission pattern of LED, and currently cover the emission cone. Figure 5.11a, b shows the cross-sectional views of detected light intensity distributions of the lensed and unlensed LEDs operating at 90 mA. All plots have been normalized with respect to the detected maximum intensity at Z = 0. The results clearly reveal that the lens array can redirect and concentrate the emitted light into the normal direction.

To examine the emission characterization of lensed LED, the ray-trace simulation on the lensed LED is performed. The stimulation model is constructed based on the experimental flip-chip lensed LED, with identical configurations as well as the material properties. The stimulated emission wavelength is set to be 440 nm. Instead of building an entire device with more than a hundred thousand lens elements, the model of the lens-integrated flip-chip LEDs is simplified and scaled down to 3 × 200 μm² so as to reduce the computational load. Half views of the

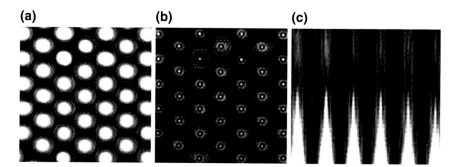

(a)   **(b)**   **(c)**

Fig. 5.10 Confocal images with captured area of 5.11 × 5.11 μm² along the x–y plane captured at **a** the sapphire surface and **b** the focal plane of lenses; **c** cross-sectional view (5.11 × 1.09 μm² along x–z plane) of the stacked confocal slices. Reprinted from Ref. [7] with permission from AIP Publishing LLC

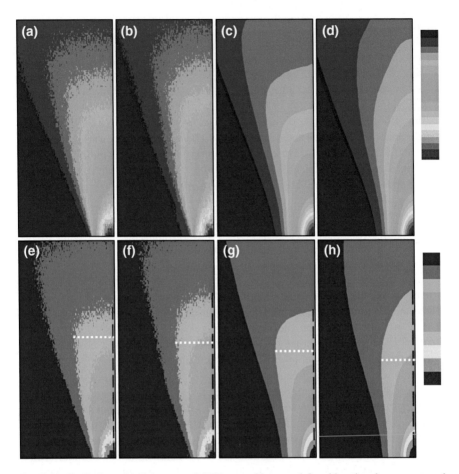

Fig. 5.11 Stacked confocal images of LEDs **a** without and **b** with micro-lens; ray-traced simulated emission patterns of **c** unlensed and **d** lensed LEDs. Both measured and simulated results are replotted into **e–h** contour plots with ten-level intervals. Dimension of the plots corresponds to the captured view of 650 μm × 1000 μm along x–z plane

computed light distributions of LEDs are plotted to have a clearl comparison with the experimental data, as shown in Fig. 5.11c, d. Clearly, the light intensities of lensed LED narrow the emission cone and significantly reduce the divergence of emitted light, as evident from Fig. 5.11d. To quantify the emission behaviors of the measured and computed results, the counter maps of those emission patterns presented by ten intensity levels are replotted in Fig. 5.11e–h. The outermost height (H) and width (W) at the 20 % intensity boundary are defined, as indicated by red solid lines and white dashed lines. The approximate values of Hs and Ws extracted from Fig. 5.11 are listed in Table 5.1. The Ws of lensed LED are reduced by 11.1 and 15.5 %, while the Hs of lensed LED are increased by 11.7 and 12.9 % for the experimental and simulation results. The simulated data provide a consistent fit to

Table 5.1 The outermost widths and heights recorded at the 20 % intensity profiles shown in Fig. 5.11e–h

	Experimental lensed LED	Experimental unlensed LED	Simulated lensed LED	Simulated unlensed LED
W (μm)	280.2	249.2	267.2	225.7
H (μm)	680.9	760.3	684.8	773.1

the measured intensity profiles. Small derivation may be due to the line and point defects which intrinsically appeared during the self-assembly of the spheres.

Without doubt, the optical behavior of micron-scale lens can be simply explained by the law of refraction and the light rays passing through lens-shaped sapphire/air interface bends. However, as the diameter of lens approaches to sub-wavelength scale, a more complex situation emerges. The subwavelength lenses have strong interactions with light and diffraction at the lens edges and interference of waves within the lens become significant [8], resulting in dramatic diminution of converging behavior. Therefore, the choice of lens dimensions becomes a critical issue in designing the optical divergence of LED emission [9].

5.6   Chapter Summary

The fabrication of a hexagonal close-packed micro-lens array has been demonstrated by the NSL approach. The flip-chip bonded LED integrated with lens array emits 27.8 % more light than the as-grown LED, attributed to the reduced total internal reflections. The lensed LED not only maintains electrical properties and enhances light extraction efficiency, but also reduces the emission divergence angle. Simulations based on RCWA algorithm and FDTD method also testify to the effectiveness of HCP lenses.

References

1. Ho JK, Jong CS, Chiu CC, Huang CN, Shih KK, Chen LC, Chen FR, Kai JJ (1999) Low-resistance ohmic contacts to p-type GaN achieved by the oxidation of Ni/Au films. J Appl Phys 86(8):4491–4497. doi:10.1063/1.371392
2. Nakayama H, Hacke P, Khan MRH, Detchprohm T, Hiramatsu K, Sawaki N (1996) Electrical transport properties of p-GaN. Jpn J Appl Phys Part 2 35(3A):L282-L284. doi:10.1143/Jjap.35.L282
3. Shchekin OB, Epler JE, Trottier TA, Margalith T, Steigerwald DA, Holcomb MO, Martin PS, Krames MR (2006) High performance thin-film flip-chip InGaN-GaN light-emitting diodes. Appl Phys Lett 89(7). Artn 071109. doi:10.1063/1.2337007

4. Chang SJ, Chang CS, Su YK, Lee CT, Chen WS, Shen CF, Hsu YP, Shei SC, Lo HM (2005) Nitride-based flip-chip ITO LEDs. IEEE T Adv Packag 28(2):273–277. doi:10.1109/Tadvp. 2005.846941

5. Choi HW, Liu C, Gu E, McConnell G, Girkin JM, Watson IM, Dawson MD (2004) GaN micro-light-emitting diode arrays with monolithically integrated sapphire microlenses. Appl Phys Lett 84(13):2253–2255. doi:10.1063/1.1690876

6. Oder TN, Shakya J, Lin JY, Jiang HX (2003) Nitride microlens arrays for blue and ultraviolet wavelength applications. Appl Phys Lett 82(21):3692–3694. doi:10.1063/1.1579872

7. Li KH, Feng C, Choi HW (2014) Analysis of micro-lens integrated flip-chip InGaN light-emitting diodes by confocal microscopy. Appl Phys Lett 104(5). Artn 051107. doi:10. 1063/1.4863925

8. Lee JY, Hong BH, Kim WY, Min SK, Kim Y, Jouravlev MV, Bose R, Kim KS, Hwang IC, Kaufman LJ, Wong CW, Kim P, Kim KS (2009) Near-field focusing and magnification through self-assembled nanoscale spherical lenses. Nature 460(7254):498–501. doi:10.1038/ Nature08173

9. Zhang Q, Li KH, Choi HW (2012) InGaN light-emitting diodes with indium-tin-oxide sub-micron lenses patterned by nanosphere lithography. Appl Phys Lett 100(6). Artn 061120. doi:10.1063/1.3684505

Chapter 6
Optical and Thermal Analyses of Thin-Film Hexagonal Micro-Mesh Light-Emitting Diodes

Abstract In this chapter, vertical thin-film light-emitting diodes with integrated micro-mesh arrays are presented. By removing the sapphire substrate through laser lift-off, vertical current conduction becomes possible, improving current spreading capability and thus electrical properties. Compared with the as-grown device, laterally guided light is capable of escaping into free space through the etched micro-mesh, evidence of which is provided by confocal imaging. At high driving current, more pronounced enhancement is observed, attributed to low junction temperatures due to efficient heat conduction as verified by infrared thermometric imaging.

6.1 Introduction

The III-nitride family of semiconductor alloys, with band gap energies spanning the ultraviolet to visible ranges, has firmly established the role of LEDs for solid-state lighting amongst other uses. While InGaN LED device structures are mostly grown on sapphire substrates, the sapphire layer imposes limitations to the performance of LEDs governed by their quantum and light extraction efficiencies. The large lattice mismatch between sapphire and GaN gives rise to high densities of dislocations within the GaN epilayers, [1] which serves as non-radiative recombination centers. The low thermal conductivity of sapphire impedes vertical heat conduction, so that the high junction temperature induced by heat accumulation reduces internal quantum efficiencies. Optically, light which has propagated into the sapphire is likely to be trapped and absorbed due to total internal reflections. Being an electrical insulator, devices grown on sapphire cannot conduct vertically, so that additional chip space and fabrication steps are required for exposing the n-GaN and to form n-contacts; such lateral-conducting devices suffer from current spreading and thus have poor emission uniformities.

With advances in laser processing, it is now possible to create vertical thin-film LED devices through laser lift-off (LLO) processes [2, 3]. Such devices exhibit

© Springer-Verlag Berlin Heidelberg 2016   93
K.H. Li, *Nanostructuring for Nitride Light-Emitting Diodes
and Optical Cavities*, Springer Theses, DOI 10.1007/978-3-662-48609-2_6

enhanced performance, benefiting from significantly improved heat-sinking and current spreading. However, a large portion of emitted photons still remain trapped within the GaN layer due to total internal reflection at the GaN/air interface, so that light extraction remains a major limiting factor in thin-film LEDs. Different strategies introduced in Chap. 2 have been proposed to promote out-coupling of laterally guided modes into leaky modes from GaN-on-sapphire LEDs. In this work, we combine a hexagonal micro-mesh array with LLO technology to build an ultra-efficient thin-film LED for optimal light extraction and heat dissipation.

6.2   Experimental Details

The epitaxial structures of the blue LEDs in this work are grown by MOCVD on c-plane crystalline sapphire. 200 nm-thick ITO is deposited as a p-type current spreading layer. Figure 6.1a shows a schematic diagram of the proposed thin-film hexagonal micro-mesh LED (TF-μLED), which consists of 15 × 14 interconnected hexagonal elements with 24.5 μm long edges separated by etched trenches of 2.5 μm in width. The hexagons are not completely isolated; instead they are interconnected via 6 μm wide bridges to allow for lateral current flow, as illustrated in Fig. 6.2. The chip with Al mirror coating on top of the ITO is mounted p-face

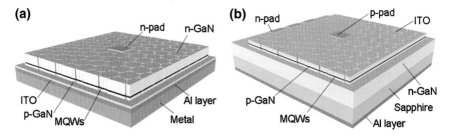

Fig. 6.1 Schematic diagrams depicting the key components of **a** a thin-film vertical micro-mesh LED and **b** a GaN-on-sapphire lateral micro-mesh LED

Fig. 6.2 Mask pattern of micro-mesh array

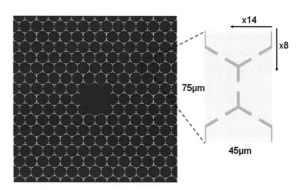

downwards onto a Cu sub-mount; its sapphire substrate is detached by LLO with the collimated 266-nm beam from an Nd:YAG laser (Continuum Surelite). To assist with separation, the sample is heated on a hotplate to melt the Ga droplets formed at the GaN/sapphire interface. The detached GaN film is immersed into dilute HCl for removal of residual droplets. The undoped GaN layer is etched away to expose the n-GaN surface by ICP etching using BCl_3/He gas mixtures. The exposed surfaces after LLO are characterized by AFM. The laterally-interconnected micro-mesh patterns, spanning a $630 \times 600 \ \mu m^2$ region, are photolithographically defined and ICP-etched through the GaN layers (including QWs) down to the ITO layer. This is followed by deposition of Ti/Al/Ti/Au n-electrodes which are subsequently annealed for ohmic contact formation. Thin-film LEDs without the micro-mesh (TF-LED) of equal dimensions are fabricated for comparison. Additionally, GaN-on-sapphire laterally-conducting LEDs of equal emission areas, with (Lat-μLED) and without micro-meshes, (Lat-LED) are fabricated. After photolithographic patterning, the top ITO layer is removed by wet-etching, followed by removal of $\sim 0.5 \ \mu m$ of GaN by ICP etching, ensuring that the micro-mesh structure has penetrated down to the quantum wells. The bottom sapphire surfaces are coated with an Al mirror to reflect light upwards. Figure 6.1b shows a schematic diagram of a Lat-μLED, in comparison to the TF-μLED.

During the LLO process, GaN decomposes into gaseous N_2 and Ga droplets at the interface. The $60 \times 60 \ \mu m^2$ AFM image in Fig. 6.3a depicts Ga droplets randomly distributed across the exposed GaN surface. After removal of droplets, the root-mean-squared (RMS) surface roughness of the surface, as illustrated in Fig. 6.3b, is evaluated as 23.6 nm. Figure 6.3c indicates that the n-GaN exposing deep etch process preserves the optically smooth surface, with an RMS roughness of 26.8 nm or $\sim \lambda/17$. Figure 6.4a–d shows CCD-captured images of the packaged devices with two major observations: (i) in Fig. 6.4b, d enhanced light intensities are detected at the etched regions within the micro-mesh, and (ii) for the lateral LEDs in Fig. 6.4a, b light is weakly emitted from the facets of the thick sapphire layer.

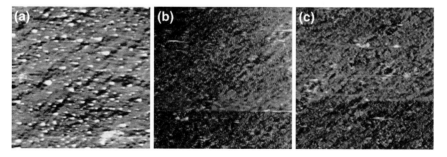

Fig. 6.3 AFM images showing surface morphologies of the exposed GaN surfaces after **a** LLO, **b** removal of droplets, and **c** ICP deep etching

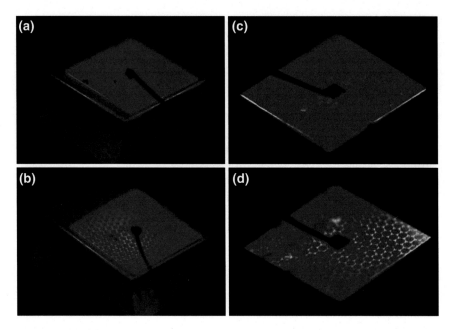

Fig. 6.4 CCD-captured images of the **a** Lat-LED, **b** Lat-μLED, **c** TF-LED, and **d** TF-μLED operated at a bias current of 10 mA

6.3   Electrical Characteristics

The I–V curves plotted in Fig. 6.5 indicate that the I–V characteristics of vertical LEDs have steeper slopes (which represent dynamic resistances) than those of the lateral LEDs. Evidently vertical LEDs exhibit lower resistances and deliver higher currents at the same bias voltage, consuming less power. The forward voltages at 20 mA are 3.48, 3.59, 3.06, and 3.13 V for the Lat-LED, Lat-μLED, TF-LED, and TF-μLED respectively. As illustrated in Fig. 6.6, the electrically insulating sapphire substrate forces the injected current to flow laterally from anode to cathode, representing a lengthy conduction path giving rise to emission nonuniformities. On the other hand, the current flows vertically in thin-film LEDs overcoming current crowding effects [4], where the conduction path is reduced to several microns giving a lower resistance, and current injection is uniform across the entire emission region, as described in Fig. 6.7. The extracted dynamic resistances are 19.82, 23.84, 10.54, and 13.14 Ω for the Lat-LED, Lat-μLED, TF-LED, and TF-μLED respectively. Reduced ohmic contact resistance may also have contributed to the reduced resistances for the thin-film LEDs; the n-contacts are formed on the etched n-GaN surface, which contains N-vacancies in the near-surface region. Due to the donor

Fig. 6.5 I–V characteristics
of fabricated devices

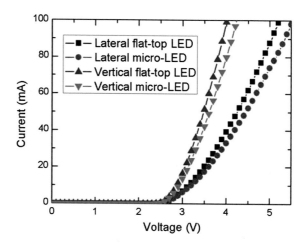

Fig. 6.6 Schematic diagrams
illustrating the current flowing
pathways of lateral LED

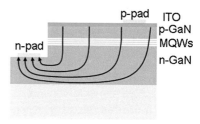

Fig. 6.7 Schematic diagrams
illustrating the current flowing
pathways of vertical LED

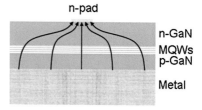

nature of N-vacancies in GaN, the surface becomes n+ giving rise to a reduction in
contact resistance [5]. On the other hand the μLEDs generally have higher dynamic
resistances than the flat-top counterparts, which can be attributed to the reduced
effective contact areas caused by surface micro-structuring.

6.4 Optical Properties

For optical measurements, the packaged and un-encapsulated devices are placed into a 12-in. integrating sphere coupled to an Ocean Optics HR2000 spectrometer via an optical fiber. The L-I characteristics plotted in Fig. 6.8 show that significant enhancement of light output powers is observed from thin-film vertical LEDs. Figure 6.10 plots light output power as a function of current normalized with respect to the Lat-LED. At a low injection current of 20 mA, the output powers of the Lat-μLED, TF-LED, and TF-μLED are enhanced by 29, 39, and 61 % respectively. The thin-film architecture avoids trapping of light in the thick sapphire substrate and absorption by the ITO layer. As injection current increases, the thin-film vertical LEDs exhibit increasing optical enhancement, while the enhancement factor for the lateral Lat-μLED maintains at the level of ∼1.3. The rate at which light output increases with current for GaN-on-sapphire devices drops rapidly, mainly attributed to reduction of quantum efficiencies induced by the poor heat conduction capability of sapphire. Removal of the sapphire substrate in thin-film devices solves this problem effectively. The integration of a micro-mesh further promotes light extraction by allowing laterally propagating photons to escape through the sidewalls of the mesh [6]. At a current of 100 mA, the emission intensity from a TF-μLED is double that of a Lat-LED. Direct evidence of enhanced light extraction is provided by the emission profiles from the LEDs are captured with a Carl Zeiss LSM700 confocal microscope using 50× (NA = 0.8) and 20× (NA = 0.5) objective lenses.

Figure 6.9a shows a 100 μm × 100 μm 3-D confocal emission image of the hexagonal micro-mesh array, where data on the z-axis represent intensity of light emitted at that particular point. The collected 3-D emission profile is an exact reverse image of the surface morphology of the micro-mesh, indicating that maximum light emission occurs at the regions between the micro-hexagonal elements;

Fig. 6.8 Plot of light-output power as a function of injection current

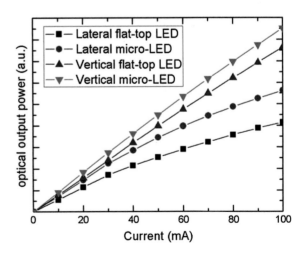

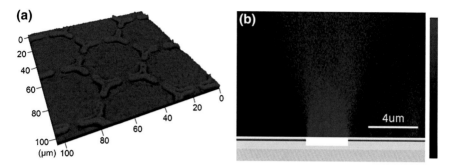

Fig. 6.9 a 3-D and **b** cross-sectional emission profiles captured by confocal microscope using 20× and 50× objective lens respectively

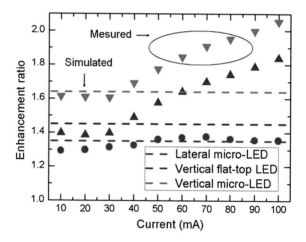

Fig. 6.10 Measured (*symbols*) and simulated (*dashed line*) enhancement ratios normalized with respect to Lat-LED

indeed light is readily extracted through the micro-mesh sidewalls. A close-up cross-sectional confocal diagram is shown in Fig. 6.9b, depicting a distinct emission cone from the region between micro-elements, constituted by emission from the sidewalls on both sides. Confocal imaging thus reveals the mechanism of enhanced light extraction from μLEDs [7, 8].

To make further sense of the measured results, optical ray-trace simulations are carried out using Tracepro. Figure 6.11a, b depicts the 3-D models of the lateral and vertical μ-LEDs adopting identical structures, optical parameters, and dimensions as the fabricated devices, except that the metal contacts and bond wires are excluded. According to the simulated results, the Lat-μLED, TF-LED, and TF-μLED emit 36, 45, and 64 % more light than a Lat-LED (dashed horizontal lines in Fig. 6.10). The analytical predictions are indeed consistent with our experimental data at low currents (<20 mA) when heating effects are negligible, since ray-trace simulations do not take thermal effects into consideration.

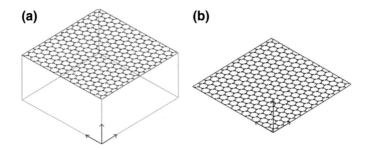

Fig. 6.11 Solid models of **a** Lat-μLED and **b** TF-μLED for ray-trace simulation

6.5 Thermal Analyses

With increasing currents, the heat dissipation capabilities of LEDs become an important factor affecting their efficiencies. To accurately probe the junction temperature distribution, the turned-on LEDs are imaged with an LWIR camera (FLIR SC645) providing 640 × 480 resolutions and an accuracy of ±2 °C. Since the measured surface is in close proximity to the junction, the measured reading is a good indication of the actual junction temperature. Figure 6.12a, b are captured emissivity-uncorrected LWIR images of the Lat-μLED and the TF-μLED operated at currents of 100 mA for 5 min. Although the temperatures in the ITO region of the chip in Fig. 6.12a appear lower than those of the surrounding GaN areas, this is simply due to the difference in emissivities between the two materials. After correction using emissivity values of $\varepsilon_{GaN} = 0.82$ and $\varepsilon_{ITO} = 0.51$ [9], the temperature distribution across the chip surface becomes homogeneous at equilibrium. For the TF-μLED without an ITO layer, the readings are taken across the chip; the data points are indicated as crosses in the figure. For consistency, the data points are taken along the edges of the Lat-μLED chip.

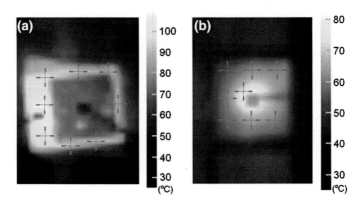

Fig. 6.12 Images of **a** Lat-μLED and **b** TF-μLED captured by infrared camera at a bias current of 100 mA

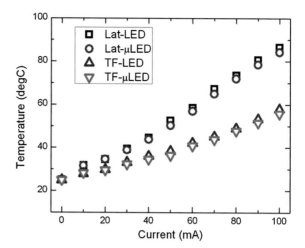

Fig. 6.13 Plot of averaged, measured temperatures as a function of driving currents

Figure 6.13 plots emissivity-corrected temperatures (averaged over all data points) as a function of currents after the device temperatures have reached steady states (>5 min). Clearly, the junction temperatures of the lateral LEDs rise more rapidly than those of the vertical LEDs. The cooler operation of vertical LEDs can be attributed to various factors. First, when current flows through a resistive device, Joule heating is inevitable. Vertical LEDs offer lower resistances due to significant shortening of current conduction paths, thereby reducing resistive heating which helps to lower junction temperatures. Second, multiple internal reflections of emitted light may eventually be reabsorbed by GaN as phonons, resulting in a rise in lattice temperatures. Nevertheless, the major reason is that heat generated from GaN-on-sapphire LEDs tends to be accumulated within the GaN layers due to the poor thermal conductivity of sapphire substrates. The sapphire-free thin-film configuration allows direct heat conduction through to the Cu sub-mount which acts as a heat sink, thus efficiently promoting heat dissipation. Although sapphire removal contributes most significantly to the lowered junction temperatures, the micro-structured surface also plays a supporting role. Since the additional sidewalls allow more light to escape, the likelihood of reabsorption is diminished, reducing heat generation. The enlarged surface area to volume ratio may also promote dissipation of heat to the ambient. Of course the higher dynamic resistances of the μLEDs give rise to more pronounced Joule heat effects.

While heat is continuously generated in an operating LED and ambient air is well-known to be a thermal insulator, the sub-mount becomes the major path for heat dissipation. To investigate the thermal properties of LED chips from another perspective, 3-D simulation of heat distribution is performed based on the finite element method. The thermal resistance across the length of a material can be expressed as $R_{TH} = L/(Ak)$, where L is the thickness of material, A the

Fig. 6.14 Simulated
temperature distributions
across the **a** GaN-on-sapphire
LED and **b** thin-film LED
under the same heat input
power

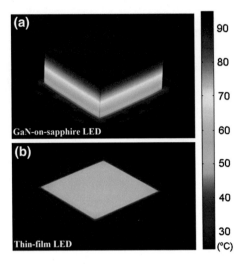

cross-sectional area perpendicular to heat flow and k its thermal conductivity. The thin ITO and Al layers have been omitted. A conductive epoxy (2.5 W/mK) is sandwiched between the chip and its Cu sub-mount (400 W/mK) while the GaN layers (130 W/mK) in both the vertical and lateral LEDs serve as the heat source. By applying an arbitrary input power of 2.2 W, the lateral LED attains much higher temperatures compared with the vertical LED, as shown by the 3-D simulated temperature profiles in Fig. 6.14a, b. The result clearly shows that the highest temperature is observed at the GaN layer and gradually decreases beyond the sapphire substrate. On the other hand, the vertical thin-film configuration allows the heat to be effectively transferred to the metallic sub-mount, resulting in significantly lower surface temperatures.

6.6 Chapter Summary

The important roles of light extraction and heat dissipation to the efficiencies of InGaN LEDs have been highlighted in this work, through lateral and vertical device configurations with and without micro-mesh arrays. The proposed thin-film vertical µLED is demonstrated as a feasible solution toward high-efficiency light generation through minimization of the current conduction path between anode and cathode resulting in reduced dynamic resistance, superior heat-sinking capabilities drastically reducing the junction temperature together with enhanced light extraction using a micro-mesh surface. At an injection current of 100 mA, the light outputs of TF-µLEDs are enhanced by a factor of two over those of conventional Lat-LEDs.

References

1. Reshchikov MA, Morkoc H (2005) Luminescence properties of defects in GaN. J Appl Phys 97 (6). Artn 061301. doi:10.1063/1.1868059.
2. Wong WS, Sands T, Cheung NW, Kneissl M, Bour DP, Mei P, Romano LT, Johnson NM (1999) Fabrication of thin-film InGaN light-emitting diode membranes by laser lift-off. Appl Phys Lett 75(10):1360–1362. doi:10.1063/1.124693
3. Kelly MK, Vaudo RP, Phanse VM, Gorgens L, Ambacher O, Stutzmann M (1999) Large free-standing GaN substrates by hydride vapor phase epitaxy and laser-induced liftoff. Jpn J Appl Phys Part 2 38(3A):L217–L219. doi:10.1143/Jjap.38.L217
4. Guo X, Schubert EF (2001) Current crowding and optical saturation effects in GaInN/GaN light-emitting diodes grown on insulating substrates. Appl Phys Lett 78(21):3337–3339. doi:10.1063/1.1372359
5. Chua SJ, Choi HW, Zhang J, Li P (2001) Vacancy effects on plasma-induced damage to n-type GaN. Phys Rev B 64 (20). Artn 205302. doi:10.1103/Physrevb.64.205302
6. Choi HW, Jeon CW, Dawson MD (2004) InGaN microring light-emitting diodes. IEEE Photonic Tech L 16(1):33–35. doi:10.1109/Lpt.2003.818903
7. Choi HW, Jeon CW, Dawson MD, Edwards PR, Martin RW, Tripathy S (2003) Mechanism of enhanced light output efficiency in InGaN-based microlight emitting diodes. J Appl Phys 93 (10):5978–5982. doi:10.1063/1.1567803
8. Griffin C, Gu E, Choi HW, Jeon CW, Girkin JM, Dawson MD, McConnell G (2005) Beam divergence measurements of InGaN/GaN micro-array light-emitting diodes using confocal microscopy. Appl Phys Lett 86(4). Artn 041111. doi:10.1103/Physrevb.64.205302
9. Chang KS, Yang SC, Kim JY, Kook MH, Ryu SY, Choi HY, Kim GH (2012) Precise temperature mapping of GaN-based LEDs by quantitative infrared micro-thermography. Sensors-Basel 12(4):4648–4660. doi:10.3390/S120404648

Chapter 7
Conclusion and Future Work

7.1 Conclusion

In this thesis, nanosphere lithography has been demonstrated as a practical alternative approach towards large-scale nanofabrication. Through modified NSL processes, the intrinsic restriction of self-assembled close-packed pattern can be overcome, to achieve diverse nanostructures which have been applied to III-nitride materials for realizing enhancement of light extraction of LED. Moreover, two sphere-patterned cavities have been demonstrated, which support WG modes and the sizes of cavities are sufficiently small to achieve high Q and single mode lasing for short wavelength emission. Here we summarize the works as follows.

In the first part, NSL has been applied to III-nitride materials for realizing enhancement of light extraction of LED. Close-packed photonic crystal structures have been successfully patterned on the ITO film of LEDs and also constructed into intermediate layers of devices to form embedded PhC configuration. Moreover, the dimensions of spheres in an array are adjusted without altering their pitch via a mild etching step, so as to achieve PBG structures. Both air-spaced and clover-shaped structures provide the tunable PBGs which overlap with emission band of LED, thus efficiently suppressing the unwanted guided mode and significantly promoting the light extraction.

In the second part, sphere-patterned cavities have been successfully applied to nitride-based materials for realizing short-wavelength cavities. Optically pumped lasing is observed at room temperature from both cavities. The nano-ring array functioned as optical resonators supporting whispering gallery modes, achieving blue light lasing with high Q-factor and low threshold. Another lasing is also observed from metal-clad pillar array, with single mode UV emission. The optical properties of the resonant cavities are further verified with FDTD simulations, which are found to correlate well with the experimental results.

Additionally, close-packed lens structure is developed on the sapphire surface of flip-chip LED. Without degrading the electrical properties, the lens-integrated

© Springer-Verlag Berlin Heidelberg 2016
K.H. Li, *Nanostructuring for Nitride Light-Emitting Diodes
and Optical Cavities*, Springer Theses, DOI 10.1007/978-3-662-48609-2_7

structures not only enhance the light extraction, but also significantly reduce the emission divergence. Far-field and near-field emission behaviors are investigated by angular-resolved EL measurement and confocal microscopy imaging respectively.

Finally, the micro-mesh array has been integrated into the thin-film LEDs via optical lithography process. Optically, interconnected microstructures provide additional sidewalls for light escaping into free space. Thin-film vertical configuration minimizes the current conduction path to promote electrical characteristics of devices. Their superior thermal properties have also been demonstrated and verified by infrared thermometric imaging.

7.2 Future Work

As demonstrated in this thesis, the self-assembled NSL process can be modified to introduce diversely optical nanostructure. By adding various micro-fabrication techniques, such as dry etching and metal coating processes, the restriction of close-packed patterning can be overcome, which potentially extends the functions of sphere-patterned structures. It is worthwhile to further investigate the self-aligned fabrication methods to produce the novel nanostructures to boost the performance of optoelectronic devices.

In Chap. 3, PBG structures patterned by modified NSL are extremely useful for boosting light extraction efficiency. To realize electrical-injected LED devices, additional processes are required to integrate the n- and p-type electrodes on those PL structures. Regrowth of a p-cap demonstrated in Sect. 3.2 is one of the possible solutions. Another alternative approach that can be pursued is to employ the spin-on-glass (SOG) technique to fill the exposed voids so as to electrically isolate the p- and n-type materials. The SOG step can also passivate sidewall defects induced during the dry etching process, but the resultant gap-filled periodic structure indeed alters the properties of PBG. Further optimization on designing the photonic band gap is necessitated so as to ensure the interaction with the emission band of LED (Fig. 7.1).

As seen so far, electrically-injected nitride cavities are not yet widely available in the marketplace. NSL has been demonstrated as a promising fabrication approach to pattern whispering gallery mode resonator because of its small size and potential for high-density integration on a wafer. The main challenge in developing GaN electrically-injected WG mode cavities is to integrate n- and p-pads on the device since direct coating of metal layers will cause electrical shortage. Although spin-on-glass technique has been developed to fill the gaps, such fabrication ambient may affect the optical properties at the boundary of WG mode cavity, thus introducing the cavity loss. Alternative approach is to regrown the top GaN layer by ELOG method so as to provide planarized surfaces without filling the voids. Further development on electrically-injected lasing operation based on NSL will certainly prove to be a great step forward in advancing nitride laser technologies.

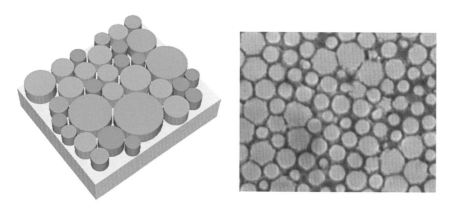

Fig. 7.1 Schematic diagram and FE–SEM image illustrating nanopillars with a wide range of dimensions

Nanospheres and microspheres, with an unlimited range of dimensions, offer the possibilities of developing 2-D HCP nanopillar arrays. Apart from improving light extraction, nanopillar can effectively boost the IQE of high-indium InGaN LEDs by relaxing their unwanted strains. By integrating the nanopillar array, the fundamental observation of spectral blue-shift indirectly indicates how much strain can be relaxed. Furthermore, nanopillars of a continuum of dimension can be also patterned by NSL, making use of a nanosphere colloid containing sphere with a wide range of diameters. With the right mix of nanopillar dimensions, different shades of light can be generated from a single chip. Most importantly, this kind of strain-relaxed nanostructures offer both high IQE and EQE, potentially making them the ideal single-chip phosphor-free solution.

Printed in the United States
By Bookmasters